D1124581

Patents for Power

Robert M. Farley and Davida H. Isaacs

PATENTS FOR POWER

Intellectual Property and the
Diffusion of Military Technology

The University of Chicago Press · Chicago and London

The University of Chicago Press, Chicago 60637
The University of Chicago Press, Ltd., London
Published 2020
Printed in the United States of America

29 28 27 26 25 24 23 22 21 20   1 2 3 4 5

ISBN-13: 978-0-226-71652-7 (cloth)
ISBN-13: 978-0-226-71666-4 (e-book)
DOI: https://doi.org/10.7208/chicago/9780226716664.001.0001

Library of Congress Cataloging-in-Publication Data

Names: Farley, Robert M., author. | Isaacs, Davida H., author.
Title: Patents for power : intellectual property and the diffusion
of military technology / Robert M. Farley, Davida H. Isaacs.
Description: Chicago : University of Chicago Press, 2020. |
Includes bibliographical references and index.
Identifiers: LCCN 2020029778 | ISBN 9780226716527 (cloth) |
ISBN 9780226716664 (ebook)
Subjects: LCSH: Intellectual property—United States. | Military-
industrial complex—United States. | Business intelligence—
United States. | Technology transfer—United States.
Classification: LCC KF2980 .F37 2020 | DDC 346.7304/86—dc23
LC record available at https://lccn.loc.gov/2020029778

♾ This paper meets the requirements of ANSI/NISO Z39.48-1992
(Permanence of Paper).

To those who led, Michael and Marilyn;
and to those who follow, Elisha and Miriam

CONTENTS

1: INTRODUCTION

Vignette: The Search for a New Rifle

After World War II, both the Soviet Union and the United States sought "assault" rifles: infantry guns that could operate effectively in single-shot, semiautomatic, and automatic modes. The desire for such a weapon came from the shared experience of fighting the German Wehrmacht, which indicated that soldiers inflicted most casualties at short range.[1] At the same time, both the Red Army and the US Army saw an advantage in a weapon that would allow soldiers to engage at the longer ranges traditionally associated with rifles. The problem for designers in both countries was to manage the tradeoffs between accuracy, hitting power, and speed in producing a gun that could become the standard infantry weapon of a mass army.[2]

The Soviet military-industrial complex eventually produced the AK-47 family of weapons, known colloquially as "Kalashnikovs," while the United States settled on the M16 family. In the intervening decades, upwards of ninety countries may have produced more than one hundred million rifles similar to the AK-47, while only three countries have produced fourteen million or so M16 originals and variants.[3] What explains the contrast in the extent of diffusion between the two weapons? Traditionally, political scientists have explained such variation primarily as a function of some combination of domestic organizational sophistication, external security need, and prestige. These factors surely had an impact, but understanding the range of diffusion requires an appreciation of the legal environments under which inventors, firms, and governments created and exported the weapons. Specifically, the difference between the AK-47 and the M16 depends, to great extent, on differences in intellectual property protection (IPP) between the United States and the Soviet Union.

The Red Army developed the AK-47 through a cooperative/competitive design process.[4] Beginning in 1946, several teams of engineers developed alternative models based on a set of Red Army requirements. During the process, teams borrowed liberally from each other's designs in order to produce the most effective weapon. This effort resulted in a weapon containing elements of many different designs, which consequently could not be fully attributed to the work of a single individual or team.[5] Despite the collaborative nature of the final design, for political reasons the Soviet government decided to credit the engineer and former Red Army soldier Mikhail Kalashnikov with "inventing" the rifle.[6] Although Kalashnikov enjoyed significant rewards from this credit, he held no property rights in the weapon or in any of its components. The government's decision to award credit to Kalashnikov led to controversy down the road, however, as other participants argued that Kalashnikov had copied design elements from various teams.

This focus on mass production over scrupulous attention to individual intellectual achievement was emblematic of the Soviet defense industrial base, which consisted of a vast network of state-owned firms producing massive amounts of military equipment for the USSR and its allies and satellites around the world. These firms conducted internal research and development, and also worked with a constellation of state-operated laboratories and research facilities on high-technology systems. Firms (and teams within firms) competed against one another to develop military technology to government specifications, with success resulting in political favor, prestige, and greater resources.[7]

Internationally, the USSR had a licensing system that granted rights of construction to allied countries; but political factors, rather than the interest of individual firms, drove technology transfer decisions.[8] In 1958 production of AK-47s began in North Korea, in 1959 in the People's Republic of China, and in the early 1960s in Egypt, all either Soviet allies or bulwarks against Western expansion.[9] Eventually, every member of the Warsaw Pact adopted the AK-47 as its primary infantry weapon, and developed its own production facilities. While most often the Soviet Union directly provided most of the technology to other countries, in other cases countries received the technology from earlier recipients, without direct Soviet coordination. For example, Albania received the basic AK-47 technologies from China.[10] Consequently, in addition to the huge stockpile of AK-47s produced by the USSR, Soviet allies and partners produced vast numbers of rifles. Monitoring production of AK-47s was made more difficult by the phenomenon of unlicensed "craft" shops around the world, which produce an uncertain number of weapons of varying quality.[11] As a result, as many as ninety countries may, at one time or another, have seen production of the AK-47 and its variants within their borders, but it is impossible to know exactly how many.[12]

The United States also sought a new assault rifle after the war. After a series of abortive efforts, the US Army issued contracts to three firms to pro-

duce the M14, a large rifle using a 7.62mm round. The M14 went into production in 1959, and served as the primary US rifle during the early years of
the Vietnam War. However, performance issues—including weight and suitability to a jungle climate—forced the Department of Defense to reconsider
the contract. An earlier call for proposals had resulted in the Armalite AR-15,
a lighter assault rifle using a 5.56mm round.[13] Colt, a major small arms production firm, acquired a license to produce the AR-15 in 1959. It used political connections in the US Air Force to good effect in opening to the door to
major US procurement.[14] Transfer of the rifle to the Army of the Republic of
Vietnam (ARVN) allowed Colt and its allies to argue for the gun's effectiveness in combat conditions.[15]

The M16 would become the primary rifle of the US Army during the latter
portion of the Vietnam War, although substantial teething problems limited its effectiveness.[16] Like the AK-47, the M16 spread across the world, becoming the primary firearm for most NATO countries and many other US
allies.[17] In contrast to the 100 million AK-47s and spin-offs, however, only
about 14 million M16 variants have been produced since the 1960s.[18] While
a wide variety of manufacturers produce variants of the M16, most are still
operating under licensing agreements with Colt, with current manufacturing only in the United States, Canada, and China. In addition to the licensed
versions, the only known unlicensed spin-off is the Chinese CQ assault rifle,
produced by NORINCO—a company that also produces unlicensed copies
of the AK-47.[19]

So why, then, is the AK-47 seemingly much more popular than the M16?
Several factors explain its greater extent of diffusion, including timing of
market entry, the configuration of Cold War alliances, and the ease of production. Nevertheless, an appreciation of the intellectual property environments helps illuminate the differences that influence diffusion: the original
development of the weapon, its early manufacturing, and finally the presence and frequency of reproductions.

During the Cold War, the United States and the Soviet Union adopted
much different models for developing new military technology. In the
United States, private industry met military demand, albeit with state support and in close collusion with military practitioners. In the Soviet Union,
the state owned the major defense enterprises. Ownership of technological
innovation or creative works followed these broader approaches: American
defense companies retained rights to their innovations, whereas the Soviet
Union recognized no independent private ownership of innovations in the
military sphere.

This Soviet absence of private ownership translated into an absence of obstacles for any manufacturer seeking to fabricate their own version of the AK-47. Whereas a business wanting to produce a copy of the M16 had to contend with the powerful intellectual property laws of the United States as well as the still more powerful Colt Corporation, which had owned the contract since the gun's inception, a company wanting to produce its own line of Kalashnikovs merely had to acquire either the plans or a working example. While the former procedure is a model for rigorous quality control and oversight of what is, after all, an engine of lethal force, the latter is one for widespread mass production under an array of conditions.

In short, the laws that structure how the inventors, producers, and distributors of weapons do their jobs have an impact on the development and diffusion of military technology. Intellectual property law, the body of law that regulates the protection of invention, has an effect on how states develop and disseminate military technology.

Introduction

Why do states adopt some weapons and not others? How does a defense industrial complex reliably produce innovative new systems? These questions have bedeviled both policy makers and political scientists for a very long time. Since the dawn of the modern state system, governments have made innovation in military technology a bureaucratic focus, bringing to bear capital, scientific expertise, and the attention of private industry in order to seek advantage over potential foes. Today, many national governments sit atop a bewilderingly large and complex system for military innovation, development, and procurement.

During the Cold War, and especially in its latter half, scholars in political science, organizational theory, and other disciplines devoted an increasing amount of attention to the question of how states pursued military innovation. In the United States these studies won government support, often through research grants and access to qualitative and quantitative data about innovation. Employing a range of analytical tools, scholars developed theories of military innovation that concentrated on civil/military relations, the nature of the international threat environment, and the internal organizational dynamics of military organizations.

While many of these studies have investigated the domestic conditions necessary for military technological innovation, few have focused on how the legal environment affects how firms and states produce these innovations, or on how international law affects the diffusion of technology across the international system. In part, these lacunae stem from deep disciplinary divides that have separated the study of international law from the study of international security. Scholars of military innovation simply do not concentrate on legal factors, while scholars of international law rarely dive deep into how military organizations make procurement decisions. This has left a hole at the center of the study of military innovation and diffusion, which is curious given that both military organizations and industrial defense producers find themselves bound by law. This study seeks to fill part of that hole by examining two related groups of questions.

Questions

First, how does intellectual property law affect the process of technological innovation in the military sphere? How have intellectual property (IP) law and the diffusion of military technology interacted and structured each other in the past? Do particular configurations of IP law speed or slow technological innovation? How does IP law affect the relationship between the state and private firms? How have technological changes interacted with domestic systems of IP protection?

Second, how does intellectual property law affect the diffusion of military technology across the international system? What limits do domestic configurations of IP place on the transfer and acquisition of military technology? How has the increasing "value added" of intellectual property changed the conduct of international espionage? How has the emerging architecture of international intellectual property law affected the trade in military technology? What impact has this architecture had on how firms organize their multinational activities? Finally, how has international IP law affected domestic systems of IP protection?

In this study we employ a mixture of analytical methods to examine the impact of intellectual property law on the innovation and diffusion of military technology. We describe the historical and contemporary contexts under which states and firms have developed military technology, both in the United States and abroad. We compare the intellectual property components of several national systems of innovation, and explore how these systems have changed over time. We investigate the role that intellectual property law and broader intellectual property concerns have played in the arms trade, and particularly in the system of export control developed by the United States during and after the Cold War. Finally, we examine how the means and methods of international espionage have changed in response to symbiotic changes in technology and IP law.

Communities

Just as this book draws on multiple literatures, it seeks to appeal to several different communities. First, this is a work of international relations theory, designed to add to the literature on military innovation, the diffusion of military technology, and the influence of law on how states make security decisions. Each of these rich literatures contain extensive empirical and theoretical work, and this study seeks to explore the connections between the three.

Second, this work hopes to contribute to the policy debates associated with military procurement and defense innovation. The production and procurement of advanced defense technology remains a core interest of the state. This study contributes to our policy understanding of the relationship between innovation, diffusion, and intellectual property law, and hopefully opens the way for the development of better decision-making processes, and the improvement of regulation associated with arms procurement and export.

Third, this book seeks to illuminate the legal complications of intellectual property law in the defense industrial complex. Law firms have increasingly become interested in the protection of intellectual property in the national security sphere, and scholars of law have produced considerable, if disparate, work on the subject. This book draws much of this work together to provide a better understanding of how intellectual property protection works in the defense industry, and how the law can contribute to defense innovation.

Finally, we hope that general students of national security, law, and defense technology find something of interest in this work. We have attempted to convey the relevance of intellectual property protection to some of the best-known accounts of the development of military technology. Our argument is that intellectual property law and the production of military technology are best understood in conjunction with one another, and that accounts which ignore either are at best incomplete. The renewed focus that the Trump administration has placed on the protection of intellectual property, especially with regard to the relationship with China, has moved IP protection to the center of the national security conversation.

Notes on Terminology

As this book speaks to several different audiences, defining several terms of art will add clarity to the discussion.

- A *national innovation system* (NIS) consists of three parts: state-controlled research agencies, research labs and educational institutions, and private firms. Together, these three groups provide the financial capital, human capital, and physical infrastructure of military technological innovation.
- The *defense industrial base* (DIB) consists of the state and privately owned firms that operate primarily in the military and defense market.
- The *military-industrial complex* (MIC) is a constellation of actors that includes the defense industrial base, the uniformed military services, and the civilian bureaucracy of defense procurement.
- *Intellectual property* (IP) consists of patents, trade secrets, copyright, and trademarks, protected under legal regimes of the state and of international organizations.
- This book uses the term *firm* in the classic economic sense, to refer to business organizations such as publicly-held corporations, state-owned enterprises, and privately-owned commercial groups.
- This book uses the term *state* in the international relations sense, to refer to national (rather than subnational) governments.

Why Should Intellectual Property Matter Now for Military Technology?

Why concentrate on intellectual property law? Historically, the security studies subfield has paid little attention to development in international or domestic intellectual property law, beyond a comment here or there about how technological innovation requires sound legal support.[20] Few studies have concentrated on how the legal regulatory regime affects a military-industrial complex, or how the emerging international intellectual property regime might affect the diffusion of military technology.[21] However, concurrent developments in intellectual property law and in the nature of the defense industry suggest that the intersection of the two demands greater attention.

First, the global defense industry depends more on foreign sales than ever before. The monopsony (single buyer) conditions that existed for most of the twentieth century no longer hold, as foreign buyers have represented a steadily increasing share of the industrial output since the end of the Cold War. Declining defense budgets in Europe and Russia have forced arms manufacturers to focus heavily on foreign customers. With reliance on exports comes concern over export controls and technology transfer, and especially concern over how foreign buyers will use, trade, and respect the intellectual property of arms producers. Even European and Russian companies have begun to recognize their need to safeguard their innovations and trade secrets in order to remain competitively viable.

Second, partnerships between firms have become ever more important to the production and development of military technology. Many of the best-known military systems of the last decade have emerged from joint ventures between large, established producers and small, nontraditional firms.[22] These partnerships inherently create the potential for conflicts over the ownership of technology and trade secrets, both between the firms themselves and between the firms and the government. To the extent that the nature of intellectual property law affects the profitability prospects of these alliances, IP law has an impact on military innovation.

Third, the defense industry has undergone both horizontal and vertical integrations that have diversified supply chains across numerous countries. A modern jet fighter contains components built in dozens of countries, often by hundreds of subcontractors. This new reality has forced US defense contractors to manage a variety of local intellectual property regimes, and this has had an impact on the manufacture of military equipment and the transfer of technology.

Fourth, the political demands of defense contracting have grown steadily more complex. Arms export agreements have increasingly taken on the character of transnational public-private partnerships. To sell fighter jets, for example, US firms might have to agree to build components in the customer country, as well as to transfer technology associated with the jet's weapon systems. This kind of commercial complexity creates the need for intricate legal arrangements that delineate where and how firms can transfer technology.

Fifth, the dual (military and civilian) use of numerous twenty-first-century technologies, especially those from within the computing and communications sectors, has complicated the relationships that private firms have with governments. While most US firms are pleased to sell to the US government, few nontraditional providers envision the Defense Department as their only customer. Rather, they plan to sell their technological innovations commercially, for civilian purposes—hence, "dual use." However, state interest in acquiring the IP rights to the data, patents, and trade secrets of firms developing dual-use technologies stands in tension with those firms' commercial efforts.

Sixth, the defense industry faces the same competition as firms in other sectors in the developed world. Firms in developing countries often have better access to local markets, and can take advantage of less rigorous local labor and environmental laws to lower their costs and shorten their production schedules. With foreign firms having clear advantages in production and labor cost, US industry in general needs to protect the area in which it maintains a competitive advantage: intellectual property and innovation. This phenomenon has provided the impetus for many large corporations in the developed world (and especially in the United States) to press for stronger intellectual property protection in global trade forums.

Seventh, private firms have, over the past five decades, contributed an increasing percentage of total funds devoted to technological research and development. At the same time, the US government remains deeply involved in most defense-oriented research, committing substantial resources to the development of new weapon technologies. Consequently, most new military technology involves some mix of public and private funding, and emerges from a mix of public and private institutions (research universities, for example). This creates problems for the ownership of the intellectual property associated with technological innovations, problems that intellectual property law can either solve or exacerbate.

Taken together, these trends point to two conclusions. First, scholarship on military innovation and diffusion has overlooked the role played by intellectual property law in the makeup of the defense industrial complex. Second, IP law potentially has a greater impact on the development and transfer of military technology than ever before. This combination of absence of exploration and increased importance suggests that a full examination of the interaction of IP law and national security is sorely overdue.

Why This Matters: Offset Strategies and the Diffusion of Military Technology

The impact of intellectual property protection on the innovation and diffusion of military technology may prove deeply consequential for strategic competition in the twenty-first century. Concerns about the innovation and diffusion of military technology have preoccupied US defense thinking since the 1950s. The United States Department of Defense has pursued three "offsets" since the end of World War II. It identified Soviet numerical superiority and the continuing wartime mobilization of the Soviet defense industry base as key advantages that threatened Western security; technology would "offset" those advantages.

The first offset, launched in the 1950s, was designed to preserve US (and NATO) technological superiority in the face of what was expected to be overwhelming Soviet conventional military superiority.[23] It concentrated on nuclear weapons and the systems needed to deliver them, though it also included conventional military aspects.[24] This first offset carried both domestic and international implications for intellectual property law. On the domestic side, it deepened the research and production relationship between the big defense firms and the government, creating the "military-industrial complex" that President Dwight Eisenhower memorably evoked in his farewell address. This deepening led to a cozy relationship between private firms and the state, relaxing the need for strict application of intellectual property law. On the international side, the first offset was accompanied by government imposition of strict export controls on weapons and military technology, which private firms accepted because they expected access to strong markets domestically, and to NATO allies.

The second offset, launched in the mid-1970s, was based on the precepts of the "revolution in military affairs" or "military technical revolution," even if the practitioners had not yet settled on terminology. It involved pursuit of technologies associated with communications, computing, and long-range precision guidance, with the goal of creating the military capability to disrupt Warsaw Pact forces in echelon as they attacked NATO's central front. The proponents of the second offset recognized the need for focusing on dual-use technologies, as civilian firms had in many cases displayed enough flexibility and innovation to move ahead of their counterparts in the DIB. In combination with the judicious use of government research funding and the promise of a significantly larger and more procurement-oriented defense budget, the second offset would maintain the position of the old defense firms while providing incentive for new contributors. It would result, its progenitors hoped, in the capacity of NATO forces to destroy the nerve centers of the Warsaw Pact, halt a Soviet advance in its tracks, and induce the disintegration of the Red Army and its Eastern European satellite partners.

The second offset also had domestic and international implications for intellectual property law. On the domestic side, the complexity of introducing new players into the defense game required the development and enforcement of new rules guaranteeing that every firm stood a chance to make a profit. This sat uneasily with the existing defense firms but, as they still controlled the overwhelming bulk of a huge domestic and international market, they could accept some restrictions in order to maintain their positions. On the international side, the US government stepped up its efforts to control the export of technology to the Warsaw Pact, clamping down on the transfer of dual-use technologies even on behalf of US allies. These controls were intended to starve the Soviet Union of the "disruptive" technologies that characterized a private-public partnership, and which the state-controlled defense industry of the USSR could not produce. The rules imposed by the United States represented a de facto international regime for the control of intellectual property, as most of the NATO and major non-NATO allies (including Japan and South Korea) subscribed to the restrictions.

The third offset, launched in 2014, seeks to preserve the superiority of US arms in the context of the economic and technological growth of key competitors, most notably the People's Republic of China. The third offset, much like the previous two, seeks to leverage the innovativeness of the US civilian economy to develop military technologies that significantly exceed the capabilities of their Chinese, Russian, and Indian counterparts. The authors of the third offset seek to achieve this by focusing even more on dual-use technologies, and by enhancing the ability of nontraditional defense providers to access the Department of Defense. This represents a far greater threat to traditional providers than the second offset, as it comes during a period of US defense budget austerity, and in the expectation that Chinese economic growth will continue to outpace American. The third offset also assumes that the traditional means of securing technological superiority, export controls, will fall by the wayside as the digitization of knowledge and the technology transfer associated with globalization of the world economy grant competitors easy access to even the most cutting-edge systems.[25]

And so, one of the core questions this study hopes to answer, or at least contribute to, is "How does intellectual property law matter for the third offset?"

Theory

This section begins to develop the theory of how intellectual property law affects the production, diffusion and innovation of military technology. This question touches upon three major debates in international relations theory: how states innovate militarily, how technological innovations spread across the international system, and how international law affects the behavior and internal constitution of states. This section summarizes the extant debates on these topics, before offering a way forward for thinking about how the three questions interact.

Innovation and Diffusion

Over the past four decades, a rich literature on military innovation and diffusion has developed in political science. Military innovation involves the development of new doctrines, forms of organization, and technologies for the defense sphere. Military diffusion involves the spread of those innovations across the international system. Diffusion necessarily comes after innovation, though innovations can also emerge through interaction between states. This section examines the most significant theoretical trends on the diffusion and innovation of military technology. We pay particular attention to the ways in which major approaches in the field treat law generally and intellectual property law specifically, in both domestic and international applications.

THEORIES OF MILITARY INNOVATION

Innovative military technologies help states to accomplish their goals, up to and including survival. States with advanced military technologies can defend themselves against aggressors, or take advantage of weaker, less advanced opponents. Accordingly, both the realist and constructivist schools of thought in international relations theory have tackled the problem of military technological innovation in some detail. This section looks at how theories of military innovation have developed as a consequence of interaction between academic and policymaking communities.

A central theme in the academic debate over military innovation pits the role of ideational (or human, in colloquial terms) factors against the role of technology.[26] Most scholarship accepts that ideas and material technology interact in the innovation of military doctrine and practice, though scholars differ on the relative weights of these factors.[27] On the ideational side, scholars have accepted a loose distinction between broad social factors (ethnic composition, state capacity, national culture) and factors specific to military practice (organizational learning and experimentation, organizational culture, bureaucratic politics).[28] Stephen Rosen, for example, discusses the role of internal military debate in the development of new tactics and strategies.[29] Colin Gray emphasizes the importance of intellectual developments in the history of warfare and innovation.[30] Generally speaking, the literature treats technological innovation as a partially exogenous variable, with states having only limited control over improvements in military equipment. The direction of technological innovation, thus, is at least partly subject to social and organizational factors, just as technology provides the context for social and organizational change.[31] The debate over the balance between technology and "doctrine"—usually defined as including experience, training, and organization—extends into the policy literature as well.[32]

As this study concentrates on the technological side, it will ignore many of the more vexing questions associated with the sources of military doctrine. However, to the extent that military doctrine and culture inform a military organization's approach to technology, they can help define what the state expects from its national innovation system (NIS) and what demands it can make upon that system. With this in mind, this study concentrates on technological rather than doctrinal or organizational innovation.

THE OFFENSE-DEFENSE BALANCE

Beginning with George Quester, a large scholarship emerged on what became known as the "offense-defense balance."[33] This literature sought to ferret out how the prevalence of certain kinds of weapon systems affected the international politics. Specific technologies, or combinations of technologies, could reduce or increase the costs of military action. For example, a common account of twentieth-century warfare ran thus: machine guns (leading to trench warfare) produced defensive stalemates, while tanks and aircraft (enabling mobile warfare) made territorial conquest easy. Arms control efforts in the early century, notably the Geneva Disarmament Conference of 1932, tried to distinguish between "offensive" and "defensive" weapons, allowing diffusion of the latter while restricting development of the former, as a means to reducing international conflict by raising its costs and lowering the probability of its success.[34]

Much of the earlier work associated with the offense-defense balance focused on the effects of the balance rather than the process of achieving it. A representative argument suggested that conditions of defensive dominance could facilitate arms control and the development of international regimes, while offensive dominance encouraged preemption and aggressive behavior.[35] Mild reformulations of the argument suggested that offensive dominance (or the perception of such) set the stage for alliance behavior prior to the two world wars.

But for several reasons, the offense-defense literature reached a dead end.[36] First, critics correctly noted that circumstances had a huge impact on whether particular technologies had offensive or defensive characteristics. The technological conditions of World War I and World War II each spurred wide variance in offensive and defensive tactics. Moreover, the success of offensive and defensive operations often depended more on doctrine and force employment than on technology.[37] Offense-defense balance theory generally lacked a good account of international diffusion (how offensive and defensive technologies spread between countries), leaving it unclear whether the balance represented a systemic variable or simply the relationship between a pair of states.

Most important, from the point of view of this study, the offense-defense balance literature as it developed in the 1980s and 1990s tended to treat technological change as an exogenous variable. As we shall see, this understanding of technology as a "pusher"—exogenous technological change spurring change in policy—mirrored much early work on technology in the field of economics.[38] Few studies within the literature examined how states pursued innovation, or under what conditions the balance might change. For the most part, states simply adapted or failed to adapt to new technological conditions, echoing classic realist thinking on state behavior. Misunderstanding the current nature of the offense-defense balance could prove disastrous to both status quo and revisionist nations, but neither had much opportunity to change the balance.

However, some source material for offense-defense theory took seriously the prospect for international law to affect innovation and diffusion of military technology. The Geneva Disarmament Conference of 1932 tried to establish legal limitation on the development of weapons with offensive functions, such as submarines, heavy bombers, and aircraft carriers.[39] The conference disintegrated not just because of the difficulty of distinguishing between offensive and defensive weapons (submarines, for example, clearly perform both roles), but also because each participant interpreted the strategic impact of offense and defense differently.[40] Many in the United Kingdom, for example, believed that deterrence against a continental power required the Royal Air Force to field a fleet of heavy bombers.[41] Despite these difficulties, some modern arms control theory also takes seriously the prospect of distinguishing between offensive and defensive weapons for purposes of arms control.[42]

REVOLUTIONS IN MILITARY AFFAIRS

By contrast, the revolution in military affairs (RMA) literature closely examined the question of military technological innovation. Stemming in part from Soviet military theory, it viewed military history as a succession of technical revolutions, each of which established the basic conditions under which states would innovate and military organizations fight.[43] For example, the revolution that preceded World War I created the "empty battlefield," as machine guns, smokeless powder, and heavy artillery forced fundamental changes in how armies would fight. Similar revolutions overtook naval warfare (the development of steel-hulled battleships, and later the aircraft carrier), and air warfare (the transition from piston- to jet-engine aircraft). Most RMA literature foresaw the dawn of a new revolution in military affairs in the 1980s and 1990s, in which the combination of highly sophisticated sensors, long-range precision munitions, and real-time communications would transform how military organizations operated.[44]

Critics of RMA theory pointed out the difficulty of distinguishing between evolutionary and revolutionary changes in military technology. Did the invention of the tank and the airplane, for example, constitute a revolution, or did they simply supplement the existing technological situation in 1914?[45] Critics also noted the context-dependence of much technological innovation. Transformations in the technological foundations of the US military-industrial complex did not, for example, make the US Army more capable of solving the counterinsurgency problem in Iraq and Afghanistan.[46]

Like offense-defense theory, the RMA literature treated much technological development as exogenous. However, RMA theory carried a more practical bent than offense-defense theory, as many of the theorists were also practitioners. Consequently, it tried to frame particular innovations within a technological context, and to point the process of innovation in specific directions.[47] Both Soviet and Western theorists of RMA, for example, argued for industrial and research strategies focused on the next revolution. Especially on the Soviet side, theorists also lamented when a nation's industrial and innovative capacity could not produce sufficiently advanced technology. And like offense-defense theorists, RMA theorists often struggled to explain international diffusion, or more specifically the lack of diffusion.[48] Revolutionary developments, especially in the latest RMA, often failed to spread beyond a few core states.[49] RMA theory had little to say about the legal framework of innovation, at either the national or the international level, beyond suggesting that economic systems which fostered and protected innovation were generally more effective at pursuing cutting-edge technologies.[50]

Nevertheless, RMA theory gives theorists and practitioners a much better handle on the direction and process of innovation than offense-defense theory. It helped open a fruitful field for considering cross-fertilization between military doctrine, the defense industrial base, and civilian technological frontiers. RMA practitioner/theorists, including Andrew Marshall of the Office of Net Assessment, played important roles in charting the direction for US technological innovation in the 1980s and 1990s.[51]

NATIONAL INNOVATION SYSTEMS

The national innovation system literature attempted to remedy many of the problems associated with previous theories of innovation by focusing on process rather than outcome. This literature does not provide an alternative explanation for innovation, but rather offers a framework for thinking about the necessary conditions under which states conduct innovative activity. A national innovation system consists of three parts: state-controlled research agencies, research labs and educational institutions, and private firms. Together, these three groups provide the financial capital, human capital, and physical infrastructure of military technological innovation.[52]

As Paul Bracken, Linda Brandt, and Stuart E. Johnson describe it, a defense innovation system (DIS) is the subset of the NIS that deals with military innovations. In the United States, "this includes institutional actors such as DOD and the services, defense contractors, and supporting institutions such as universities. Investment flows of research and development (R&D), Wall Street valuation of defense contractors, and the amount of dual use are all part of the defense innovation system."[53] A key component of the DIS is the research, development, and acquisition system (RDA), which is the bureaucracy for managing the development and production of weapons.[54] According to Tai Ming Cheung, the "defense RDA apparatus refers to the complex ecosystem of organizations and rules responsible for the conceptualization, design, engineering, testing, production, and operation of weapons systems.[55]

As a framework for developing theories, the NIS certainly lacks the parsimony and elegance of the broader theories of military innovation. It has the advantage, however, of giving us a productive glimpse into the process of innovation, and into the decisions that characterize how, whether, and when states will pursue technological innovation. By focusing on the activities of individual firms and agencies, it also proves compatible with theories on the changing nature of the firm, and on the changing preferences of government entities.

However, the NIS literature unquestionably leaves open considerable space for discussion of the legal framework of technological innovation generally, and the intellectual property framework specifically. Establishing that an NIS exists as a constellation of private firms, public corporations, educational institutions, and government agencies inevitably raises question about the linkages among the elements of the constellation, and about the rules that regulate their interaction. The NIS literature is largely compatible with the literature on military-technical revolutions; indeed, practitioners have used the two in conjunction to productively examine the proliferation of dual-use technologies in military equipment.

This study uses the NIS model as a starting point for considering the impact of intellectual property law on military innovation. In chapter 3, the study evaluates how the US national innovation system uses intellectual property law to manage relations between its components. Chapter 4 does the same thing for several comparative systems, including the Russian, Chinese, and South Korean.

Theories of Diffusion of Military Technology

The international diffusion of military technology is related to, but distinct from, the issue of innovation. Generally speaking, the literature on military innovation treats it as a domestic or even intraorganizational process. By contrast, the study of diffusion involves evaluating how innovations developed in one state spread to other states. The relationship between the two is important but often underspecified, as the systemic effects of military innovation—changes in the nature of warfare, and modified political arrangements based on those changes—depend less on how innovations come about than on how they spread between states.

The literature on diffusion in military affairs focuses on three questions. As characterized in Emily Goldman and Leslie Eliason's edited volume *The Diffusion of Military Technology and Doctrine,*

> The first debate concerns how one defines the diffusion process, which is critical for identifying whether or not diffusion has occurred. The key question here is whether the communication of information is sufficient to conclude that diffusion has taken place. . . . The second debate concerns the causes of diffusion. What motivates states to adopt innovations from abroad, and what is the mechanism by which knowledge is transferred? While scholars advance various typologies, three distinct processes— competition, socialization, and coercion—drive the spread of policies across societies with different implications for what is modeled. The third debate concerns the patterns and effects of diffusion.[56]

Our argument mostly concerns the second and third questions, on the motivations and mechanisms for—and through which—states acquire knowledge, and on the overall impact of legal regimes on the patterns and effects of diffusion. The rest of this section details how the extant theories of military diffusion have approached these questions, once again with an eye to how they treat international and domestic law as plausible influences on the process. Two primary mechanisms for the diffusion of military technology are the international arms trade and industrial espionage. Broadly speaking, two schools have emerged regarding why states pursue certain military technologies. The first, associated with structural realism, suggests that states pursue technologies they believe to be effective because they want to survive. The second, associated with more sociological approaches, maintains that states pursue technologies because of a complex combination of material and normative factors.

REALIST-RATIONAL SCHOOL

Kenneth Waltz's spare account of international politics argues that anarchy, the ordering principle of the international system, forces states to compete with one another in order to survive.[57] One strategy states use to survive in an anarchic system is to mirror other the behavior of other states; as a result, they need to pay attention to the strategies pursued by other states so that they do not fall behind. Because of the importance of military doctrine, Waltz gives it special attention. He argued in his landmark work *Theory of International Politics*:

> The fate of each state depends on its responses to what other states do. The possibility that conflict will be conducted by force leads to competition in the arts and the instruments of force. Competition produces a tendency toward the sameness of competitors. Thus Bismarck's startling victories over Austria in 1866 and over France in 1870 quickly led the major continental powers to imitate the Prussian military staff system. . . . Contending states imitate the military innovation contrived by the country of greatest capability and ingenuity. And so the weapons of major contenders, and even their strategies, begin to look much the same all over the world.[58]

In short, competition and the fear of state death drive the spread of military innovation. Military organizations in contending states learn from organizations in successful states, imitating strategies and innovations that prove worthwhile. Successful strategies then spread across the international system. Competitive imitation plays a crucial role in the diffusion of military technology.

According to neorealist theory, military organizations learn vicariously from each other and through scanning (paying attention to the international environment).[59] Organizations consciously imitate the successful practices they see in other organizations.[60] Competitive pressures force military organizations to match the effectiveness of other organizations.[61] Just as firms attempt to copy the successful practices of other firms in order to keep costs low, military organizations have to adopt the innovations of other organizations in order to have a reasonable chance on the battlefield.

Scholars working in the neorealist tradition have provided nuance to this argument. States adopt technologies and doctrines more readily when they feel threatened; at times of reduced threat, they may let innovation slide.[62] Joao Resende Santos argues that states adopt new doctrines, organizational modes, and technologies out of concern for their security, with adoption succeeding insofar as states can devote sufficient resources to the project.[63] In *The Diffusion of Military Power*, Michael C. Horowitz takes an organizational perspective to extend this case, arguing that differences between wealth and organizational complexity limit the diffusion of military power.

Modern military operations place enormous demands on human and financial capital.[64] Consequently, some states and societies lack the wealth and organizational capacity to successfully adopt certain kinds of military innovations. Andrea and Mauro Gilli have modified this argument to suggest that states require an entire ecosystem of related technologies in order to support certain innovations—an argument that has implications for industrial espionage.[65] The Gillis argue that the increase in technical sophistication of military technology over the twentieth century has made it more difficult for "second mover" countries to adopt the most advanced weapon systems, such as stealth aviation.[66] Along similar lines, Douglas O'Reagan argues that even great powers struggle to acquire technological when they lack access to tacit knowledge (human know-how).[67]

NORMATIVE SCHOOL

Sociological explanations for institutions and behavior presuppose that humans live within a universe of social meanings. Although interest plays a role in behavior, appropriateness and legitimacy help construct the conditions under which states interpret interest. The behavior of others, especially powerful states, legitimates some behaviors and delegitimates others.[68] Norms and expectations structure how states pursue their interests. For example, in *The Purpose of Intervention*, Martha Finnemore argued that "the utility of force is a function of its legitimacy."[69] States may behave in accordance with a utilitarian logic, but rules of appropriateness help determine the parameters of utility. Several scholars have applied this logic to procurement. Emily Goldman has explored how the impact of the Western military model differed in Japan and Ottoman Turkey, and Theo Farrell has studied the effect of world military culture on the constitution of the Irish army.[70] Particularly relevant for this study, Dana Eyer and Mark Suchman established that poor countries buy expensive weapons, even when cheap weapons would better meet their needs.[71] Similarly, Daniel W. Henk and Marin R. Rupiya argue that symbolic logic drives much procurement strategy in African states.[72]

The sociological framework allows that consequential and social logics interact, but that it is worth the effort to specify how, and under what conditions, such interaction produces varied outcomes. Ideas matter, and the presence of powerful ideational forces at the systemic level can cause states to redefine their identities and change the methods through which they pursue power. National security problems extend beyond survival. Changing ideas about the constitution of national power can, therefore, result in changes in how states conceive of power and, consequently, how they conceive of competition with one another. For example, the 1941 Japanese attack at Pearl Harbor helped change how the world thought about battleships, even though battleships remained useful weapons of war. Also, the lack of a sharp distinction between symbolic and utilitarian action may activate familiar international relations dynamics such as the security dilemma. Symbolic action always has a practical impact, and the indistinguishability of symbolic from consequential acts in the international system may lead to the perception of all such acts as consequential, with the result that states feel threatened, and act threatening in response. Moreover, symbolic interactions can take on a competitive dynamic all their own. Governments pursue prestige for both domestic and international audiences, and both pursuits can manifest in competition with other states.[73]

Dennis Gormley works through the implications of this logic on the cruise missile trade in *Missile Contagion*.[74] Gormley argues that international law has treated cruise and ballistic missiles differently, and that this treatment has had significant implications for how the two types have spread. International arms control efforts have generally concentrated on the threat of ballistic missile diffusion, implicitly treating cruise missiles as a more tactical concern. This has persisted even as the functional gap between cruise and ballistic missiles has narrowed. Gormley attributes the focus on ballistic missiles, in large part, to the disinclination of the United States to accept any restriction on the production, development, or employment of cruise missiles.[75] Given the US dependence on cruise missiles in many of its recent conflicts, this disinclination is hardly surprising, but Gormley argues that it has a global impact on the availability and desirability of cruise missiles.[76] When the United States uses cruise missiles, it normalizes them for the rest of the international community.

LAW AND INTERNATIONAL POLITICS

Law and international politics have a complicated relationship. War, or intercommunal violence, seems particularly resistant to legal regulation. As early as Cicero, warriors and statesmen claimed that the law falls silent during war.[77] As Carl von Clausewitz suggested:

> Attached to force are certain self-imposed, imperceptible limitations hardly worth mentioning, known as international law and custom, but they scarcely weaken it. Force—that is, physical force, for more force has no existence save as expressed in the state and the law—is thus the means of war; to impose our will on the enemy is its object."[78]

The development of the realist, and later neorealist, IR theory came as a scholarly rejection of the efforts to manage international affairs by legal means in the first half of the twentieth century.[79] The perceived relevance of international law goes to the core of differences between the major schools of thought on international relations. Historically, practitioners of international law have sought to constrain state behavior, while practitioners of more traditional forms of statecraft have often tried to evade or skirt these legal constraints. This divide has extended into the scholarly realm, making legal-focused studies of "hard" military questions, such as the innovation, production, and diffusion of military technology, relatively rare. This study seeks in part to bridge that divide by bringing law back to the core of "hard" security studies.

Peter Gourevitch laid out the classic theoretical formulation of how the international system affects domestic politics in "The Second Image Reversed."[80] In reference to Kenneth Waltz's categorization of theories that derive international political outcomes from the characteristics and behavior of states, Gourevitch created a framework for thinking about how the international system could affect the internal operation of states.[81] Scholars have used this framework to examine the impact of war, trade, and international law on the domestic functioning of nation-states.[82]

Few disagree that the density of law and obligation in the international system has increased. Many scholars question, however, the depth of state interest in complying with this regulation, especially as compliance becomes costly. Edward Luck has framed the problem of compliance with international law thus:

> When a state becomes party to an international convention, it undertakes certain legal obligations and, in some fashion, can be held accountable for upholding its provisions. The degree of its commitment to ensuring that the larger purposes of the convention are fulfilled in specific cases and in operational terms, however, is necessarily a political matter. There is no automaticity here, particularly given the paucity of compliance or enforcement provisions in most international conventions. The degree of commitment, moreover, will depend on to whom the legal obligations are seen to be undertaken; to people, to the world, or to the nation.[83]

In short, states sometimes enter into agreements that they have no intention or capability of fulfilling.[84] Given the anarchic nature of the international system, decisions to comply depend on political calculation between competing factions and interests. Our theories of international relations suggest that states especially prioritize maintaining their political autonomy in the security sphere; consequently, scholars of security affairs doubt that international law can have much effect on state behavior.[85]

International law scholars call the difference between commitments and behavior "the compliance gap."[86] States make commitments to international law or international institutions for a wide variety of reasons; they may want access to markets or technical knowledge, they may face alliance pressures, important domestic constituencies may want an international commitment, or larger countries may lean on them. The space between the legal requirements of commitment and actual state behavior becomes the compliance gap.[87]

International treaty law has automatic and semiautomatic mechanisms for enforcing compliance at the domestic level. Much international treaty law requires, upon ratification, the domestic execution of the legal obligations in the treaty. Self-executing treaties become domestic law as soon as states formally ratify them. Non-self-executing treaties require additional legislation.[88] In some special cases, institutional mechanisms ensure the domestic adoption of international legal agreements. In the case of the European Union, for example, international treaties and acts of the European Parliament automatically ensure a degree of compliance across the organization, though individual governments may vary in enthusiasm for enforcement.[89]

This process does not happen apolitically. The functionalist school of European integration hoped that the creation of international obligations could, through the development and linkage of national bureaucracy, create regional integration without need to resort to messy intergovernmentalism.[90] The history of the European Union has made the strong form of this view untenable; integration is more punctuated than incremental. However, the weak form survives, and offers a good explanation for the development of certain kinds of state and national interests.[91]

Yet an important element of how international law and international institutions come to have a domestic footprint, potentially one that avoids the messy national politics, is through the bureaucratization of international priorities.[92] Many international agreements require the creation or modification of state bureaucracy in order to fulfill the obligations of the agreement. Whether or not states intend to fulfill these obligations, they often develop bureaucratic institutions in order to demonstrate good faith. International obligations can also privilege certain political and bureaucratic actors at the expense of others, thus tipping the balance between factions.[93]

The best examples of the integration of international law directly into the concerns and behavior of states in the security sphere comes with the law of armed conflict (LOAC), and with a variety of different arms control treaties. The former has, through a series of treaties and agreements, effectively become customary international law, binding upon every country in the world.[94] Governments which fail to embed LOAC in the decision making of their military and security institutions run the risk of severe sanctions and the prosecution of their military and civilian leadership. Compliance with arms control treaties often requires the embedding of international legal restrictions within systems of export control, otherwise used to prevent the spread of sensitive military and dual-use technologies, or to protect intellectual property.[95]

With respect to intellectual property specifically, compliance with international IP law places obligations on all sectors of society. The state must not only avoid infringing on foreign property itself, but undertake the regulation of the behavior of private actors. Similarly, private firms need to observe the law while also reporting infringements to state authorities. However, given the overwhelming state interest in maintaining a healthy defense sector (and especially in pursuing innovative military technologies), we can expect the state to pay close attention to the regulation of IP in the national security sphere.

Increasingly, a bureaucracy for managing intellectual property has become a standard component of statehood across the international system.[96] Even states that produce little intellectual property have established such bureaucracies. Institutions that manage international intellectual property protection have actively undertaken efforts to deepen their reach into domestic law. These institutions, including the World Intellectual Property Organization, negotiate with national governments to improve the level of IP protection.[97] This effort includes the contribution of legal and technical expertise in maintaining a system of IP protection. Attracting foreign direct investment (FDI) from major multinational firms often requires such protection. In other words, international regimes (including the international IP regime) do not merely "trickle down" into the domestic institutions of their member states; they actively replicate themselves, both through direct negotiation between states and by the active machinations of the institutions themselves.

WEAPONIZED INTERDEPENDENCE

One consequence of the expansion and bureaucratization of international law may be what Henry Farrell and Abraham Newman have termed "weaponized interdependence."[98] Farrell and Newman argue that, among other phenomena, international commerce creates a complex and asymmetric map in which certain players can leverage their central positions in order to coerce others. Farrell and Newman illustrated the argument with an application to the global financial system, in which interdependence has allowed the United States to aggressively target countries and firms that have violated sanctions against Russia and Iran. While seeming obvious, the argument runs contrary to decades of thinking about the "flattening" nature of international commercial networks.

We can think of the global system of patent and trade secret protection as a similar kind of interdependent network, with its own vulnerabilities and opportunities for leveraging lethality. As we will demonstrate in chapter 2, the United States has been at the forefront of an effort to increase intellectual property protection for at least the last three decades.[99] Since the 1980s, it has facilitated the establishment of a global IP protection architecture, and ensconced that architecture in multilateral and bilateral treaty law. While the impetus for strengthening international IP protection came largely from private actors, the weapons of interdependence need not be designed with lethal effect in mind. Empirical research on globalization has long suggested that states benefit from "first mover" steps with respect to the establishment of global rules of the road.[100]

Weaponized interdependence cuts both ways. The interconnected nature of the global tech sector gave Chinese firms such as Huawei access to technologies it otherwise would not have been able to take advantage of. International intellectual property protection, a system of mutually supporting patent law and trade secret protection that enabled sharing, provided the foundation for that success. But, as we will argue at various points in this volume, interdependence has given the United States leverage for attacks against China's technology sector.

Our Expectations

Political scientists have not yet developed a model for how intellectual property law affects the innovation and diffusion of military technology, largely because they have yet to think very hard about the question. Nevertheless, we can derive a few hypotheses from the theories described above.

With respect to military innovation, political scientists of the realist school of international relations thought might generally grant that patterns of intellectual property protection have an effect on how a defense industrial base operates, but they would also expect that the pressure of maintaining a competitive military would push contending states towards similar internal arrangements.[101] In other words, if a particular system of intellectual property protection helps spur innovation in one state, other states will generally copy that system and implement it on their own.

With respect to military diffusion, theorists of international politics would likewise not deny the proposition that states take protection of their technology and intellectual property seriously. However, theorists, especially from the realist school, view states as security maximizers, and generally believe that international laws and norms breach the military sphere. In other words, few states would entrust protection of their intellectual property, or that of their domestic firms, to international law or international management. Pairs or small groups of states might engage technology transfer to facilitate an arms deal, but the transaction would remain at "arm's length," and would not involve particularly sensitive technologies. States would not, as a general rule, trust other states to protect their most advanced military innovations. Furthermore, with respect to industrial espionage, realist scholars would expect states to engage in spying in order to acquire knowledge of foreign military technologies, and hopefully to replicate the most successful technologies. Realists would not, as a general rule, expect the development of norms of restraint to guide this behavior.

By contrast, we argue that the developing constellation of international intellectual property law regimes will increasingly constrain the behavior of states, even in the defense procurement sphere. The development of international IP legal regimes has forced nearly all states to create domestic intellectual property management bureaucracies. Although these bureaucracies may initially play only a symbolic function, over time they have a greater impact on state behavior.

The model works as follows. The demands of both international organizations and domestic economic innovation require the creation of an intellectual property protection bureaucracy that can negotiate with international organizations as well as private firms. Over time this bureaucracy takes on a life of its own, changing how a government operates. States will increasingly struggle to maintain dual-track systems of IP management that protect domestic innovation while facilitating the theft of foreign IP. This system will come to characterize transaction in military and dual-use equipment, thus bringing IP law into the security sphere.

The emerging international intellectual property regime, manifested in both multilateral—agreements on trade-related aspects of intellectual property rights (TRIPS) and transatlantic trade and investment partnerships (TTIP), for example)—and bilateral agreements, affects the nature and content of intellectual property regulation at the domestic level, thus yielding changes in national innovation systems that result in a more homogenous global intellectual property regulatory regime.

In large part because of trends in technology and international trade, the innovation and production of military technology has taken on an increasingly international character. The development and assembly of specific weapons now occurs transnationally, with components built in several countries and transferred to others for final assembly.

Intellectual property regulation affects the extent and nature of military innovation. Systems of IP regulation consistent with the emerging global regime tend to enable states to take advantage of dual-use technologies and enhance public-private collaboration. These systems tend to discourage unlicensed appropriation of foreign military and dual-use technology.

The emerging system of intellectual property protection facilitates the globalization of military production by reducing concerns about the loss of IP to partners. Believing they have legal recourse in the case of violation, states and firms become more willing to engage in collaborative transnational projects, including both innovation and production. The emerging system of intellectual property protection also ameliorates some of the concerns states have over illegal appropriation in the wake of arm's-length arms transfers. While many arms transfers do not involve appropriation concerns because the buyer lacks the technological foundation to copy any relevant innovations, some transactions between technologically similar countries run the risk of appropriation and theft. The existence of a common set of intellectual property protections helps to reduce this risk.

During the Cold War, the United States developed a wide-ranging system of export controls for military and dual-use equipment. This system encompassed not just the United States but also most of its partners and allies, especially those engaged in cooperative military production and research. Today, that system has become integrated with the emerging international IP regime to place strict restrictions on how and where partner states can export military equipment.

The combination of shifts in technology and the increasing relevance of IP law, both internationally and domestically, has made the field fertile for the emergence of new forms of industrial espionage. The digitization of technology, to some extent abetted by the expansion of IP protection, has made it possible for the military and intelligence assets of some states to appropriate the military and dual-use technology of others. This comes in the form of cyber-espionage, a practice that increasingly dominates efforts to steal military industrial secrets. As with many new forms of state interaction, however, the practice of cyber-espionage takes cues from the development of international laws and norms, including those of IP protection.

The Plan of the Book

The rest of the book proceeds as follows. Each chapter begins with a short vignette about the impact of intellectual property law on the technologies of national security. These vignettes serve to familiarize the reader with the history and stakes of IP protection in the defense sector. Chapter 2 describes the nature and history of intellectual property law and its intersections with the military-industrial complex. This includes descriptions of patent, copyright, and trade secret protections (including those for patent acquisition procedures), and how each of these matter for the practices of firms within and around the MIC. Scholars in business, economics, and organizational behavior fields have studied the differential impact of different IP schemes in cross-national context. While few of these studies concentrate specifically on the defense industry (which usually exists in an idiosyncratic position with respect to IP law), they provide useful context for thinking about how variations and changes in IP law might affect innovation and productivity. The chapter traces the history of intellectual property protection, highlighting instances in which military considerations have had an important impact on how the current regime came into existence. It concludes with an account of the emergence of the international intellectual property regime, with a particular eye to how the United States and other important actors have used bilateral and multilateral agreements to create an international IPP standard.

Chapter 3 delves into the US national innovation system, describing how the existing military-industrial complex came into existence in the United States. This account concentrates on the relationship between government and private business in the defense sector, including an explanation of how the United States has primarily relied on patent and trade secret regimes to protect innovative technology. The chapter highlights two intellectual property protection issues in the current US legal regime; the Invention Secrecy Act and the state secrets privilege. These two issues structure and complicate the relationship between the US government and the private defense sector.

Chapter 4 provides an introductory description of how different, often competing, intellectual property frameworks operate, including a historical account of how states have approached the protection of intellectual property. The chapter compares how the Soviet Union, the Republic of Korea, and the People's Republic of China pursued innovation within their particular intellectual property frameworks, and investigates the benefits and drawbacks of each system. Both the Soviet Union and China concentrated their defense industrial bases in state-owned enterprises, albeit with important differences in the treatments of interfirm collaboration and intellectual property. These systems have changed over time, however, with the role of intellectual property expanding as the networks of firms, labs, and government agencies became more complex. The Republic of Korea, one of the few countries to occupy a space as both an importer and exporter of high-technology military equipment, represents a particularly interesting case of collaboration between domestic and international public and private producers.

Chapter 5 examines the interaction between the international arms trade and the growing body of domestic and international intellectual property law. The chapter begins with an extensive description of the functioning of the extant arms export market. It examines how intellectual property concerns have affected arms transfers, and how states and firms have navigated the increasingly complex defense supply chain in the context of various intellectual property law jurisdictions. In particular, the chapter details the development and functioning of the US government's system of export control during the Cold War. Although it operated on an idiosyncratic set of legal principles, this system did and does represent a form of legal intellectual property protection, limiting the transfer of technologies to states and firms unfriendly to the United States. This chapter also includes a quantitative investigation of how intellectual property practices affect arms exports.

Chapter 6 studies the phenomenon of industrial espionage, especially as practiced through cyberwarfare. Although states have long engaged in industrial espionage in the defense sector, the advent of the digital age, combined with an increase in the importance of intellectual property and changes in IP law, have transformed the field of play. States no longer need direct physical access to foreign technology in order to copy and appropriate it. Indeed, this form of espionage has come to dominate discussion of cyberwarfare between China and the United States.

The final chapter reviews the findings of the book, and charts the way forward in both policy and research terms. On the policy side, we offer suggestions for reforming elements of the Department of Defense's intellectual property practices within the United States. We also offer some input regarding how the military-industrial complex in the United States should approach multilateral negotiations on intellectual property protection. On the research side, we describe several questions that could animate future researchers, working from a variety of methodological approaches.

2: THE INTERNATIONAL RELATIONS OF INTELLECTUAL PROPERTY PROTECTION

Vignette: Patent Law and the Quest for Maritime Dominance

"Water, water, everywhere—nor any drop to drink." The despair in "The Rime of the Ancient Mariner" was based on hard facts: In the seventeenth and eighteenth centuries, the ability of ships to conduct long-range military and commercial voyages depended to a great degree on their access to fresh water.[1] Ships destined for India, China, or the Americas needed either large stores of water or access to fresh water ashore. Efforts to construct a device that could convert seawater to freshwater thus became a key scientific and engineering priority in Great Britain and the Netherlands from the seventeenth century on.

It is therefore unsurprising that these efforts triggered a series of legal disputes around attempts to obtain patents—the legal mechanism by which an inventor obtains a temporary monopoly over a new invention. The early English patent system involved a complex relationship between inventors, the Crown, and a variety of interest groups, with authority to assign ownership delegated to the Royal Society of London (one of the first chartered scientific societies). Several inventors attempted to register with that society machines that they claimed could convert seawater to freshwater, with one William Walcot succeeding in obtaining a patent that gave him a monopoly over one design.[2] Walcot eventually became embroiled in a dispute with Robert Boyle over claims to a very similar, competing machine, and lost most of his rights over the invention. The strategic significance of the legal argument becomes apparent given Walcot's response: he picked up and moved to England's primary military competitor, the Netherlands, where he was able to register his patent.[3]

As history records, none of the machines ever worked.[4] The culmination of the dispute, however, illuminates how the early modern state attempted to use intellectual property protection to facilitate the extension of military reach and national power. Even at this early point, state authorities appreciated that the structure of intellectual property law does not merely influence the broad health of the national economy, but could more directly impact the pursuit of military power.

The second half of the nineteenth century saw tremendous innovation in naval warfare, driven in great part by the patent system.[5] Steam-powered vessels had begun to appear in the first half of the century, and when steam was united with screw propulsion (propellers), it resulted in fast, efficient, well defended warships. The first such battleship, the French *Napoleon* of 1850, could outpace the older sail-driven ships of the line that had dominated warfare since the seventeenth century.[6] Iron plating came next. The French armored cruiser *Gloire*, with iron plating over a wooden hull, showed the way. The iron-hulled HMS *Warrior* entered service a year later. In two years, the American ironclads CSS *Virginia* and *USS Monitor* would fight to a draw at Hampton Roads. The age of the wooden sail-driven ship of the line had ended, but innovation had not.

The next generation of ships carried heavy steel armor and breech-loading turrets, and eschewed sails completely.[7] In the United Kingdom, the firm of Armstrong-Wentworth secured several critical patents on artillery and armor technology, then used those patents to win contracts with the Royal Navy.[8] However, even as British IP policy was successful in encouraging this modernization, the British government also recognized one fraught tightwire: How to permit useful private property protection—which the government had concluded was helpful to spur such improvements—while nonetheless ensuring government access to those improvements. As is expanded upon later in this chapter, the British government used two weights to balance the established property protection: reservation of the right to compel inventors to license their military technology, and the setting of royalty rates for the government's use of that technology.[9]

Meanwhile, the US military and its counterparts in Europe did their best to get access to the latest British inventions. By 1900, many navies around the world possessed squadrons of heavily armed and armored battleships and cruisers.[10] The modern ships' combination of armor and firepower put smaller ships at a distinct disadvantage. Further innovations in fire control helped increase the range of these larger guns. These trends drove an increase in ship size, gun power, and armor into the first decades of the twentieth century.

In several countries, inventors and engineers quickly responded by shifting to develop a counter to these tremendous ships: accurate, easily fired torpedoes that, unlike the early stationary mines, were self-propelled.[11] With sufficient range and accuracy, these torpedoes could bypass the systems of armor protection that battleships and armored cruisers relied upon.

But for lethal accuracy, they depended upon a propulsion system, a detonator, and a mechanism for maintaining depth and course,[12] all needed to withstand the rigors of combat.[13] When all these parts worked successfully, they could be expected to give small, fast, maneuverable vessels a means for destroying capital ships for the first time in decades.

This budding arms race put the existing national innovation systems under strain. First, the extent and the cost of modern warships put production beyond the capacity of private industry to independently develop updated versions.[14] As a result, the US and British governments began directly employing many of the engineers working on torpedo technology. Moreover, much of the capital necessary to facilitate research and exploit innovations came directly from the government.[15] This shared investment inevitably produced questions about who owned the resultant technology in the form of patents and trade secrets. Each of those governments would come to control many of the rights associated with the patents, including domestic and international rights of transfer. Private firms, on the other hand, came to expect that the state would finance their research activities, at least in part. The torpedo provided a template for the development, production, and sale of modern military technology—a template that would help create an infrastructure of intellectual property law in both countries, and around the world.

Left inchoately addressed, however, was a second tightwire: how to protect the state from letting militarily sensitive technology fall into problematic hands, yet still permit firms to gain as much economic benefit as possible from their inventions through export of the technology.[16] The mechanisms established for managing the development of the torpedo would have far-ranging effects on the intellectual property systems of both the United States and the United Kingdom.

Introduction

Intellectual property law is certainly older than the modern nation-state. Nascent forms of patent and trade secret protection began to emerge during the medieval period, and even then they were often in association with the need to protect monopolies or provide incentive for military innovation. But the twentieth and twenty-first centuries have seen an explosion in the extent and influence of intellectual property law. An increasing number of interest group disputes, international disagreements over enforcement, and court decisions have broadened the scope of the legal conflict and added to the body of law. These concerns and disputes have impacted how states create, transmit, and access defense technology. Yet, until very recently, there has been little interest in considering how intellectual property law affects defense industries and the transfer of defense technology between states.

To understand the interactions, it is crucial to appreciate the mechanisms of intellectual property, its genesis, and its subsequent development through the twentieth century. While political scientists have only rarely discussed intellectual property, economists and lawyers have probed questions associated with IP protection at considerable depth. This chapter examines the extant research on the effect of intellectual property law on economic behavior more broadly, before drilling down to evaluate the behavior of firms in the defense sphere. We use the unfortunately awkward term "firms in the defense sphere," rather than "defense firms," quite intentionally. As will be discussed later in more detail, in light of the expanding role of dual-use technology in the modern defense arena, a proper assessment of the influence of intellectual property law requires consideration of a broader vista not only of defense-focused firms, but also of general commercial companies whose technology is sometimes of great strategic value.

After establishing the state of research on the impact of patents, this chapter turns to the international factors facilitating the spread of patent protection across the international system. States have sought coordination on IP law since the late nineteenth century, but the effort has accelerated in the past twenty-five years. This effort, including both multilateral and bilateral components (both most often led by the United States, at least until recent shifts of policy under the Trump administration), has resulted in a global homogenization of intellectual property protection around the US model. Finally, the chapter continues to examine the intertwined nature of IP law and the pursuit of national security.

The Forms of Intellectual Property

As currently understood, there are four main types of intellectual property protections, but only a triad of them come into play throughout the course of this book: utility patents, trade secrets, and copyright.[17] When lay people refer to "a patent," they are generally referring to the majority subset more accurately described as a "utility patent"—the time-limited monopoly granted by a government to inventors of a sufficiently novel and useful process, product, or device. This contrasts with the small minority of "design patents," which protect only the nonuseful designs of useful inventions. Because the functional benefits of inventions interest military-industrial complexes, and because most of the defense industry's technology is protected through utility patents, in this book the term "patents" will refer to utility patents.[18] The very nature of utility patents is a quid pro quo; in order to receive that time-limited exclusion, an applicant must disclose to the government the detailed mechanics of an invention and explain not only how it differs from prior creations in the field, but how it is sufficiently original so that "a person of ordinary skill in the art" (referred to in some jurisdictions as "the common technician in the field") could not have developed it on her own. The government, generally, then publishes that information so that other parties can copy the version once the patent exclusion has expired and, just as importantly, can use the more general knowledge for other purposes right away. With that general knowledge, others can create related products and often even create work-arounds that accomplish the same goal in a fashion similar to the version covered by the patent, but are tweaked just enough to avoid infringing on the patent's territory.

An alternative form of protection relies on the creation of "trade secrets." Most broadly, a trade secret can be any economically advantageous information which is not generally known and is kept secure so that it is not easily ascertainable. Like Coca-Cola's well-known secret flavoring recipe "7X," trade secrets can include innovative processes and new items.[19] Any trade secrets are, by definition, *not* disclosed to the outside world—and so cannot be simultaneously patented. Instead, owners of trade secrets seek to protect them by instituting special procedures for handling them. If patent protection is akin to placing money in a bank, trade secret protection is more like keeping cash in a mattress, with the inherent benefits and risks that the metaphor implies. Like depositing assets in a bank, patent recognition involves a formal process and opens the property to government scrutiny, but comes with the assumption of certain government protections (though this assumption may turn out to be unwarranted when the state secrets privilege is unsheathed). By contrast, like keeping money in a mattress, trade secret protection is a discreet process. As the protection primarily hangs on the difficulty of uncovering private conduct, the owner attempts to maintain the confidentiality of underlying information, and often even the existence of the secret itself. But there are various avenues of remedy when that security is breached and a trade secret is stolen. While theft of a trade secret used to be matter of civil concern and financial recovery, it is a sign of the increasing importance of that approach that such theft is now frequently treated in the province of criminal prosecution. However, like the unreported money in a mattress, the trade secret legal regime provides no protection against those who, without engaging in improper acquisition, uncover the secret property either through reverse engineering of a properly acquired product or simply through independent creation.

Discussion of how intellectual property can be used to impact the dissemination of defense products would be incomplete without discussion of the related fields of design patent and copyright. Like utility patents, design patents and copyrights provide for limited-time monopoly rights. But instead of focusing on practical innovations, both design patents and copyrights are meant to protect nonfunctional design and expressive creative content. On first glance, defense technology appears unlikely to contain such creative, "artistic" aspects. After all, in the development stage, a military's first or even second priority is not its simplicity or beauty, as might be a priority for a commercial company such as Apple. However, even industrial design—that is, design of items that are primarily functional, such as automobiles and airplanes—can contain nonfunctional ornamental components that may, in combination, fall within the protection of a design patent.[20]

Readers are undoubtedly most familiar with copyright as it is used to protect works valued primarily for their original communicative content, such as movies, music, and other creative arts. But undoubtedly most significant in the defense arena is the use of copyright to protect computer software. Going forward, it is probably accurate to say that almost all technically sophisticated items will contain software. While software may contain algorithms that would satisfy the general principles of patentability, it is currently unclear under what circumstances it will be protectable by patent.[21] It is clear, however, that source code, if it satisfies the very low creativity requirement, can be copyrighted.[22]

Intellectual Property and Innovation

But what of the actual impact of IP protection on the behavior of inventors? Does intellectual proportion protection spur or hamper innovation? This question has vexed inventors, merchants, and policymakers since the beginning of the early modern period. In both the United States and the United Kingdom, policy makers and commentators have hotly debated the question of the value of patent protections.[23] While the fully formed systems of intellectual property protection eventually emerged in the US and the UK (European continental states proceeded more slowly), these systems developed fitfully, and occasionally faced strong opposition. This opposition first developed because the earliest patents were often unconnected with innovation, and existed primarily to allow the state, and particularly royalty, to reward supportive constituencies. To many in the United Kingdom, patent protection appeared to be of so little value to inventors, scientists, and industrialists that in the late nineteenth century a coalition of them argued not merely for its reform but its abolition.[24] This group argued that patent protection represented unwarranted government intrusion into the functioning of the capitalist economy. This coalition failed, although its efforts eventually succeeded in streamlining the patent process in the United Kingdom.[25]

A similar effort developed in the United States in the 1930s and 1940s. Partially in response to the Great Depression, New Deal policy makers considered options that would free scientific and industrial patents from monopoly control.[26] Although the power of the abolitionist force waned quickly, suggestions for intervention into the patent system remained until the 1950s. This included a proposal to empower a government board to distribute certain patents according to social and scientific need.[27] Another proposal was one which, if it had succeeded, would have radically transformed the nature of US defense innovation: that any government investment in a research project should void all private patent rights.[28] Criticism of the patent system also attacked the state's use of secrecy to prevent the spread of information (we will return to this in chapter 3).[29] The appropriation of a wide range of German intellectual property during and after the war gave these arguments some steam. However, even though this movement found some high-powered intellectual support (from the philosopher Michael Polanyi, among others), it failed in the long run to either abolish or fundamentally reform the handling of intellectual property in the United States.[30]

Scholarship in economics was slow to take up the question of the effectiveness of IP law in spurring innovation. For a long time, economic analysis treated technological change and invention as exogenous to the economic system.[31] The early-twentieth-century work of Jacob Schmooker helped clarify the impact of economic activity on the pursuit of innovation. Schmooker recognized that "technological progress is not an independent cause of socio-economic change, and an interpretation of history as largely the attempt of mankind to catch up to new technology is a distorted one."[32] Inventive activity, as represented by research and development investment and the establishment of patents, follows economic demand.[33] Schmooker showed this empirically by examining patenting behavior in railways and associated industries. Research within an industry, he concluded, does not "exhaust" innovation, in the sense of pushing particular industries into unproductive dead ends; rather, as long as demand exists, investment and research will follow.[34]

The landmark contribution of William Nordhaus a half-century ago tightened this weave by making inventive activity endogenous to modern economic models, particularly intellectual property protection. Nordhaus determined that, theoretically at least, firms should be sensitive to changes in patent law, especially regarding the length of patents. Through use of intellectual property protection, the state could manipulate the incentives firms had for investing in research and development.[35] Some scholars have suggested that the correlation between patent rights and innovation would not necessarily be a positive one; rather, patent rights can in certain situations dissuade innovation. Indeed, some in the Marxist tradition have argued that the commoditization of knowledge that the emergence of the international intellectual property regime represents is responsible for economic stagnation and the growth of inequality in late capitalism.[36] Refining the techniques for maintaining ownership of knowledge, by this account, curtails open research and the sharing of knowledge for the human good.

In a recent review, Mike Kimel made this argument, noting that in years where there are more patents per capita issued, the subsequent growth rate in real per capita GDP over a ten-year period seems to go down.[37] He has offered at least two reasons for this. First, wherego the creation of legal protections, so goeth attorneys fees; those additional transaction costs inevitably decrease the amount of innovation that might otherwise be created. While that is possible, we suspect that it is of very limited import, as the cost of those fees are generally going to be dwarfed by the potential profits of most defense innovations. Thus, the marginal harm to innovation caused by attorneys' fees could be expected to be quite low. However, Kimel's second point ties into wider arguments against patent protection: To a patent owner's competitors, the patent has a negative value—it creates a threat of potential damages should one's innovations fall within the scope of the patent. And that is more likely than one might imagine. Innovation, as important as it can be, is often incremental. It can be difficult if not impossible to ascertain ahead of time how broadly a patent's terms will be construed. A competitor may forgo attempting to innovate in a specific field if the result may run afoul of an invention that appears to have wide protection.

One clue as to the effect of intellectual property protection on innovation is the empirical evidence regarding how modifications to IP law have affected firms' engagement in innovative activity.[38] A growing community of scholars has tackled this question.[39] For example, Edwin L. C. Lai showed that the impact of intellectual property protection depends both on global economic position, and on the nature of technology transfer.[40] Strong IP protection in the "global North" slows innovative activity by providing Northern inventors with longer, more secure periods of monopoly control. However, strong IP protection in developing countries can help generate foreign direct investment and technology transfer, which then increases the rate of product innovation, as well as increasing the relative wage and other economic macro-indicators. On the other hand, Mariko Sakakibara and Lee Branstetter found little indication that the major Japanese patent reform of 1988, which brought Japanese practice much closer to the US standard, had a significant effect on the research and development behavior of Japanese firms.[41] Combining quantitative data with qualitative interviews, Sakakibara and Branstetter could not demonstrate that the 1988 reform changed the way in which firms approached research and development, or that it increased the overall innovative output.[42]

Data problems have slowed research. Scholars like using patent registration data because by its nature and purpose it is publicly accessible and thoroughly classified according to industrial sector and type of technology. However, drawing conclusions from patent data is problematic. A 1987 article by Bjorn Basberg laid out many of the problems and much of the debate associated with using patent registration data for measuring innovation.[43] First, not all innovations are patented. Given the public nature of patenting, some inventors may prefer secrecy (trade secrets) to protect key innovations, especially if the patents do not result in some commercially obvious technology.

Second, and perhaps most important: not all patents represent innovation. Indeed, correlating the quality of innovation by the *number* of patents is a flawed exercise. As indicated earlier, companies can be motivated to patent so as to encourage potential competitors to relinquish that field. Because it is generally not possible to fully ascertain the strength or weakness of a patent without engaging in litigation, having many patents can serve the purpose of discouraging others from spending research and development funds in that field. It might be that some of the patents would be found to be invalid or weak—that is, they would be shown to have been unentitled to patent protection. But a patent holder is aware that some potentially competing companies will choose to forgo setting themselves up for a patent battle and will instead focus their efforts on other technology. In addition, some companies often pursue even meritless patents to fool unsophisticated stock market evaluators, who will then drive up stock prices. Recognizing this, researchers have made efforts to define which patents are meaningful by focusing on particular words that likely suggest that the innovation revealed in those patents will have greater import in that technology sector.[44] Equally important is that variance in patent law and attitude toward patents may frustrate studies that attempt cross-national, cross-sectoral, or cross-temporal comparisons.[45]

In light of the expanding role of dual-use technology in the modern defense arena, a proper assessment of the influence of intellectual property law requires consideration of a broader vista. With some important exceptions (the pharmaceutical industry, for example), patent statutes themselves are not industry-specific, and the general rules of statutory interpretation should be the same across industries. Nonetheless, patent law has a variable effect on innovation across industries. Different industries have characteristics—organizational structure, means of information diffusion, or potential application of the law—that affect the impact of patent law on innovation. As Dan Burk and Mark Lemley write, "As a practical matter, it appears that while patent law is technology-neutral in theory, it is technology-specific in application."[46] Some research supports this suggestion. Albert Guangzhou Hu and Ivan P. Png posited that "as patent rights change at the national level, industries within a country may react differently according to the importance of such rights to the respective industries."[47] They found that stronger patent rights were indeed associated with faster growth among more patent-intensive industries.[48] Michael Heller and Rebecca Eisenberg, for example, specifically argued that the nature of biomedical research causes patent protection to be likely to curtail effective research within that community.[49] This book is focused on provisions of intellectual property law and policy that particularly affect military acquisition and use. As chapter 3 will discuss, the existence of the Invention Secrecy Act and the availability of government immunity among other facets of intellectual property law means that defense firms and their partners face a different legal environment than do most other private companies.

Trade Secrets

While economists have largely dominated the academic literature on patents, organizational theorists have taken a larger role on the question of trade secret protection. In part, this is because analysis of trade secrets poses obvious data problems, as academics have no more access to trade secrets, individually or in aggregate, than anyone else. It is unfortunate for there to be no empirical data regarding trade secret protection, because in a modern, interconnected economy, trade secret protection may be having a big impact on innovation. As James Pooley, deputy director general of the Innovation and Technology Sector of the World Intellectual Property Organization, argues, "Although it may seem paradoxical, trade secret laws can enable and encourage technology transfer, because they provide a commercially reasonable way to disseminate information."[50] In the absence of such laws, firms would lack confidence in their employees, customers, suppliers, and other partners, leading to deadweight economic loss and a decline in innovation. Andrea Fosfuri and Thomas Ronde, for example, demonstrated that trade secret protection tends to incentivize high-tech clustering, which itself generates economic growth and technological innovation.[51] That is because—as Silicon Valley's computing industry exemplifies—high-tech industries can benefit from labor and related capital mobility across relatively constrained geographical areas.[52] Trade secret protection enables that clustering by giving firms sufficient confidence to allow employee mobility, which in turn permits the dispersion of broader skills and knowledge gained in creating a trade-secreted product. However, the power of trade secrets is controlled both by the ability of a particular holder to effectively limit access, and by the willingness of private and state competitors to respect those attempts to limit access.[53] The former will be discussed in the context of needed defense industry collaboration; the latter becomes an issue, discussed in greater length, regarding some states' unwillingness to respect attempts at maintaining secrecy.[54]

History of Intellectual Property Acquisition as Statecraft

Why would governments, particularly capitalist ones, agree to give some person or entity exclusive rights to make or use an invention, rather than permit society to immediately benefit from the reduced costs of competition?[55] Historians trace early intellectual property systems to the Renaissance-era Italian city-states of Venice and Florence, with the perceived success in spurring innovation there leading to analogous systems elsewhere in Europe.[56] However, as those patents and copyrights required authorization from the reigning monarchs, they were often were used to grant a stream of revenue to favored individuals and groups and, in the case of copyright, to control the publication of ideas.

Over the centuries, however, the derived benefit to the public became the focus. This understanding, particularly in the case of patents, relies on the belief that an incentive is needed to get an inventor to allocate the resources to invent (given the opportunity costs involved) and to give other potential inventors access to that advancing technical knowledge through the patent protection process's detailed disclosure requirements. The patent monopoly and its associated profits are the incentives provided to achieve those two aims. There is also the tertiary public benefit that exclusivity motivates potential competitors to develop alternative designs or processes that achieve the same goal, yet work around the protected version. Often it is in creating these "work-arounds" that the technology improves even further, beginning the cycle anew. It is this public benefit rationale that explains the prefatory language in the US Constitution's clause providing Congress the right to establish copyright and patent protection: "To . . . promote the Progress of Science and use Arts."[57] This assumes that personal benefit is required in order to motivate innovation; if the inventor was otherwise motivated to equally innovate, this legal incentivizing would be economically unwarranted. This view also assumes that the public benefit from patents is greater than the cost of monopoly pricing.

Until the late nineteenth century, both the beneficiary "public" and the innovators and creators were understood to be limited to a particular state's citizenry.[58] As a result, the unauthorized appropriation of *foreign* intellectual property "was seen not only as not dishonorable, or 'piracy,' but as the rendering of a public service."[59] Indeed, today's industrial powers achieved development in part by freely appropriating the intellectual property of other countries. The British patent system was sought to lure foreign technologies to the kingdom not by granting rights to the original, often foreign, inventors, but to whoever brought inventions into domestic public knowledge. The United States' antagonism toward protecting foreign inventors was even less subtle: foreigners were ineligible to obtain US patents until 1836, and even thereafter remained saddled with higher patenting fees until the United States was forced to abandon that approach as a result of the Paris Convention.[60] Discrimination against foreigners effectively encouraged technology transfer. As David Jeremy observed, "If both citizens and aliens were denied the possibility of a patent for introducing a foreign invention, foreign inventions could be introduced to America without the additional cost of the inventor's monopoly rights. The USA therefore had access to the world's technology at a lower cost than other nations."[61] And indeed, the United States was a net technology importer throughout most of the nineteenth century.[62] One of the better known examples of appropriation was British steam engine technology, acquired with no compensation and in spite of British prohibitions on export.[63] Indeed, as late as World War II the United States was regarded even by its allies as having a loose appreciation for the intellectual property of others.[64]

Internationalization

The degree and nature of IP protection *should* vary across nations. Countries vary in terms of their exposure to international trade, the size of their domestic markets, the role of private versus public capital, the level of economic development, their trade partners, the balance of industry versus commodities, their colonial heritage, and the nature of their international commitments. All these factors have an effect on economic policy and legal systems, so we should expect that different countries in different positions will have different systems of IP protection.[65] Yet in reality, state practice in IP protection has seen an increasing degree of homogeneity.

Through the first half of the nineteenth century, efforts at creating multinational intellectual property protection arrangements ran aground in the face of pirate communities. In the United States, for example, both publishers and industrialists bitterly fought any suggestion that the US government should recognize British intellectual property.[66] The prosperity of the latter, and in many cases the survival of the former, depended on cheap access to British creative work. Senator Henry Clay took up the cause of developing a trans-Atlantic accord on intellectual property in the 1840s, but supporters of international protection failed to overcome domestic opposition.[67]

Other international efforts also fared poorly. The confederation of German states that existed prior to 1871 failed to develop a uniform system of IP protection, leading to widespread piracy—and, some have argued, considerable industrial success.[68] Not even the British Empire could manage a common standard of IP protection, as the government recognized that uneven development across the empire made patent protection onerous to the less technologically advanced colonies.[69]

Nevertheless, by the end of the nineteenth century, the "national" phase began to give way to an "international" one as US inventors joined their European counterparts in seeking to protect their inventions from unauthorized copying within the small number of industrialized countries. This phase began with bilateral deals for mutual protection of intellectual property, but soon began moving towards comprehensive multilateral treaties.[70] These international efforts resulted in the Paris Convention for the Protection of Industrial Property in 1883 (regarding patent protection), and the Berne Convention for the Protection of Literary and Artistic Works in 1886 (regarding copyright protection), in which signatory countries agreed to domestically codify certain basic protections of intellectual property.[71] The Berne and Paris Conventions formally established both priority irrespective of the nationality of the owner (permitting a foreign creator to establish priority over a subsequent domestic user) and national treatment (requiring that each nation treat foreign and domestic creators alike in establishing rights and analyzing infringement claims—so that, for example, a state may not offer its citizens longer patent or copyright terms than those provided to foreign holders). The Berne and Paris Conventions' equalizing concepts have been reaffirmed and expanded upon in the 1994 Trade-Related Aspects of Intellectual Property Rights agreement (TRIPS).[72]

However, these early agreements created only a very limited degree of legal cohesiveness and consistency. First, they permitted individual states to retain their own patent and copyright systems, specifying length of protection, scope of protection, and carve-outs permitting "fair use" and other limits on exclusivities. As Peter Drahos has noted, throughout the early twentieth century, "it was an accepted part of international commercial morality that states would design domestic intellectual property law to suit *their own* economic circumstances. States made sure that existing international intellectual property agreements gave them plenty of latitude to do so."[73] To this day, despite many industrialized countries' quite similar patent procedures, there is no such thing as an international patent. Each state typically still controls its patenting process through its own national patent office.[74] Second, and crucially, the early multilateral agreements did not contain any enforcement mechanism, making their value wholly subject to the signatories' willingness to self-police: a US patent could only be enforced by a US court, a Russian patent could only be enforced in a Russian court, a Chinese patent could only be enforced in a Chinese court, and so forth. As a result, each state was generally free to interpret its own law as narrowly as it chose.[75]

Despite nascent internationalization, the long history of wholly domestic patent enforcement frequently causes lay people to misconstrue patent law as being completely territorially limited. Many states do attempt to control inventive conduct well outside of their borders, and the United States falls unreservedly within this group. In the United States, "direct infringement" encompasses not only the creation, sale, and offer to sell an infringing product, but also the importation of that product. For example, any foreign company that ships its infringing products into the United States is a direct infringer, even if the company neither has manufacturing facilities nor marketing or sales personnel within the United States. Even a foreign company that merely induces infringement or importation can be held responsible. It is true that a foreign company may be able to avoid the damages from lawsuits outside its state of origin, as enforcement of judgments in foreign courts is difficult at best. However, many judgments can be enforced against intentional countries by seeking to reach the assets of those foreign companies through their local subsidiaries or other assets that make their way into or through the commercial or monetary system. For example, in 2019 the Trump administration determined to punish the Chinese technology firm Huawei for suspected espionage by cutting Huawei off from all US-patented technology. This has the potential to devastate Huawei's operations, as its complex technology depends not only on specific US patents, but also on patents from foreign firms that themselves depend on US technology.[76]

Changes in the late 1960s and 1970s signaled an international recognition in the mid-twentieth century of the growing importance of intellectual property to global commerce. In 1967 the formation of the World Intellectual Property Organization helped with the global administration of existing international IP agreements; in 1978, the Patent Cooperation Treaty

(PCT) entered into force, providing a unified procedure for filing patent applications in each of its contracting states, thus expediting and strengthening protection. However, that treaty has not fully unified patenting laws even in signatory countries; rather, it merely streamlines the patent process by engaging in a cooperative review that is generally accepted by state patent offices. For example, the European Patent Office does not provide any "European patent." It merely offers an expedited avenue for filing in all of the cooperative countries; an EPO grant must then be converted into individual national patents in accordance with each European state's laws.[77]

Throughout the 1970s and 1980s, US policymakers and business leaders increasingly recognized three independent but nondiscrete facts: the growing importance for the American economy of innovative technologies, and the growing importance of foreign markets for US technologies, and the increasing unauthorized appropriation of American technologies by those within developing countries, especially in Asia.[78] The United States spent the 1980s and early 1990s working with the European Union, Japan, and several other developed nations to negotiate the TRIPS agreement, which was concluded in 1994 at the end of the Uruguay Round of the General Agreement on Tariffs and Trade (GATT).

Because ratification of TRIPS is a compulsory requirement of World Trade Organization membership, any country seeking easy access to the numerous international markets opened by the WTO must enact the strict intellectual property laws mandated by TRIPS. A signatory's failure to comply with TRIPS can lead to discipline through the WTO's dispute-settlement mechanism, and accordingly through restrictions on a state's right to participate in the WTO. For this reason, TRIPS is the most important multilateral instrument for the globalization of intellectual property laws. States, like Russia and China, that were very unlikely to join the Berne Convention have found the prospect of WTO membership a powerful enticement. As a result, TRIPS established a powerful enforcement mechanism.[79]

TRIPS can affect state action to a far greater extent than earlier agreements, as it has instituted much stronger standards for IP protection, to some extent globalizing what before had largely been the province of domestic policy. However, the impetus for the TRIPS agreement originated not with the governments themselves but with multinational corporations. Pfizer, in particular, as well as IBM and other major multinational corporations with strong interests in both patents and copyrights, pushed for intellectual property harmonization. For example, the Advisory Committee on Trade Policy and Negotiation (ACTPN), which was established by the 1974 Trade Act to "institutionalize . . . business input into US trade policy and multilateral negotiations," was initially chaired by Pfizer's CEO.[80] US businesses had traditionally maintained a strong pro–free trade view, and intellectual property exclusivities, by definition, encumber free trade. However, the ACTPN and lobbying organizations controlled by these multinational organizations engaged in a systematic effort to shift the mentality of both government and major policy think tanks toward seeing intellectual property exclusivities as an aid, rather than an obstacle, to US trade growth by "relocat[ing] the intellectual property issue within a frame of fundamental liberal values—the individual right of property ownership; the right to a reward for labour; fairness."[81] They also worked to replace, in policy and news communications, the use of the more neutral terms "appropriation" and "copying" with the critical nomenclature of "piracy."

Thus, TRIPS dispenses with the justification, real or pretextual, of providing intellectual property exclusivities as an apparatus for public benefit, and transmutes that into a private property right. TRIPS does include some "public interest" exceptions which allow for compulsory licensing, but it still requires payment of even licensing royalties. However, in clear indication of which states the most powerful patent holders originated from, those TRIPS exceptions only permitted domestic production—meaning that countries without production facilities could not use compulsory licensing and contract with a supplier in another country.[82] Thus, the move to a global intellectual property system has not simply expanded the geographic scope of protection and a more effective enforcement regime, but has been part of a rapid substantive swing toward placing private commercial interests over state benefit.[83]

More recently, in the same internationalist directions, there have been moderately successful efforts to move past the purely intrastate litigation that have resulted in states' selective enforcement of patents in defense fields—though these efforts have notably lacked the full support of the defense powerhouses. The Comprehensive and Progressive Agreement for Trans-Pacific Partnership (TPP), which covers many Asian countries (other than China and Korea) representing approximately 13 percent of the world's trade, established an avenue through which a company that unsuccessfully enforced its patent rights could sue a state itself in an international arbitration forum, at which point it could argue that its loss in domestic courts had been the result of state-motivated discrimination against the foreign patent holder.[84] Originally, the United States was also a signatory in January 2016, but it withdrew from the agreement in January 2017.[85] In a January 2018 interview, US President Donald Trump announced his interest in possibly rejoining the TPP, but only if it were a "substantially better deal" for the United States. Whether this is based on a concern that the agreement would harm industry, as opposed to being based on domestic politicking, has yet to be established.

The Transatlantic Trade and Investment Partnership (TTIP), between the European Union and the United States, contained a similar provision; however, the negotiations toward that expected agreement were halted indefinitely following the 2016 US presidential election. By providing a way for private inventors to hold a state accountable when it rejects legitimate claims on national security grounds (as well as for other less legitimate reasons, such as bribery), the mechanism potentially greatly weakens a state's ability to shield itself from patent enforcement threats.[86] As a result, it is a noticeable shift in defense policy for the United States to have pulled out of these emergent agreements.

BILATERAL AGREEMENTS

In part because the multilateral institution-building process slowed around the turn of the twenty-first century, the United States and the European Union have both made intellectual property protection a key pillar in their bilateral trade agreements.[87] For example, the US-Korea free trade agreement of 2012 emphasized the protection of intellectual property rights above and beyond what multilateral institutions demanded.[88] The 2003 US-Chile free trade pact provided a template in this regard, with a full chapter devoted to intellectual property.[89] These bilateral agreements generally contain more aggressive protection of IP than is required by TRIPS, and are frequently referred to as "TRIPS-plus" agreements. One study found that twenty-five of twenty-eight such agreements contained more IP protection. The agreements also contain enforcement mechanisms similar to those of the TPP, including a dispute resolution committee and changes in the domestic law of the partner country.[90] These agreements sometimes require changes to IP legal protection in the host (US or EU) country as well.[91] The recently renegotiated United States-Mexico-Canada Agreement (USMCA), successor to the North American Free Trade Agreement (NAFTA), differs from its predecessor mainly in the extent of its IPP provisions.

Conclusion

Our understanding of how technological innovation relates to the intellectual property protection system remains incomplete. We know that patent behavior follows an economic logic; innovations happen because firms can expect a return on their research and development either through patent protection or through trade secret protection. We have a good account of why firms very likely respond to changes in patent law, though developing a good basis for how to quantify and assess those responses presents difficulties. Our theory of the relationship between trade secrets and innovation is even more difficult to study.

Whether intellectual property law reform has happened because of a better understanding of the determinants of innovation, or because powerful actors have done an increasingly effective job of maintaining monopoly control over their inventions, there is little question that the regulatory environment for managing intellectual property has changed. Moreover, this change has shaped intellectual property internationally. The next chapters take a closer look at the formal institutions by which governments (especially the United States, but also China and Russia) manage the intellectual property aspects of their military-industrial complexes. These systems of IP protection have developed against the complementary backdrop of rapid industrial globalization and the spread of IP law across the international system.

3: INTELLECTUAL PROPERTY AND THE US DEFENSE INDUSTRIAL BASE

Vignette: Working with Government

A young David with a new slingshot approaches its Goliath—not to battle, but to cooperate. In 1991, the small engineering firm Crater developed a coupling device (or "wetmate") that links underwater pipes or cables together without the need for welding.[1] Recognizing that they were having little success marketing their invention, they contacted Lucent Technologies, a subsidiary of AT&T. Lucent expressed an interest in obtaining a prototype of the so-called wetmate mechanism with the possibility of using the device to fulfill a classified contract with the US Navy. Although the project remains classified, it may have involved efforts to tap undersea communications cables.[2] Lucent obtained from Crater access to the inventors' undisclosed engineering drawings, as well as a license to use the drawings to produce a prototype for R&D purposes. In return, Lucent agreed to keep the drawings secret and to negotiate another license agreement if they were interested in using the device in their network.[3] The inventors were excited to learn that Lucent's suitability testing indicated that the coupler was the best device for the project. They were less excited, however, when Lucent offered only $100,000 to license the technology, a small fraction of the overall Navy contract. Although no licensing agreement was reached, the inventors had reason to suspect that Lucent had gone ahead and used their technology. In May 1998, Crater filed a lawsuit in federal court alleging patent infringement, misappropriation of trade secrets, and breach of contract.[4]

However, two fists—in the form of statutes known only to the most sophisticated patent attorneys—effectively smashed Crater's case. First, the inventors found their patent infringement claim dismissed by a statute shifting liability from the contractors (Lucent) to the government, and requiring that the inventors seek restitution through an administrative mechanism.[5] This effectively gutted the possible award on their patent claim. But even once the district court litigation resumed on the remaining trade secret and contract claims, Crater found itself stymied. The government acknowledged that Lucent made thirty-six Crater-based prototype wetmates for research and development, but denied that these prototypes were used in the Navy project. When Crater asked for evidence that the chosen device did not make use of their design, the government refused to produce any such information, citing the state secrets privilege. Without that evidence, Crater was unable to demonstrate violations of its trade secrets and contract claims. David, in short, was left without a rock for his sling.

Industry professionals suggest that situations like this happen with distressing frequency. As discussed in this chapter, the development of military technology increasingly became a public-private partnership, and the attribution of credit became a legal and financial issue.[6] Firms that retained rights to their intellectual property, or that at least could sell those rights for a reasonable price, prospered. But at the same time, the value of the rights themselves came into question as the military conducted testing, developed new requirements, and highlighted new techniques for innovation.[7]

Introduction

Historically, the US defense industrial base (DIB) has consisted of a varying number of large private-sector corporations. Most of these firms came into existence in the first half of the twentieth century, often in association with the aerospace sector. These firms competed for contracts and research funds from the US government, but also sought contracts with foreign governments. The US government took an active interest in the health of the DIB, sometimes taking into account the structure of the industry when tendering contracts—for example, by choosing struggling firms in order to keep them in existence.[8]

The contemporary American national innovation system (NIS) involves a constellation of government actors (including the military services, the Department of the Defense, the Department of Energy, the intelligence community, and others) and a large collection of private entities. While the most well-known private entities are those private businesses that concentrate primarily on the defense market, there are also research institutions (including public and private research labs and research universities) and various private businesses primarily focused on producing for the civilian market.

How did the defense industry become mired in intellectual property law? As discussed in chapter 2, modern intellectual property law in the United States resulted partially from the need to respond to challenges in motivating government/private collaboration.[9] Collaboration became necessary as the costs of research and development more generally began to clearly exceed the means of private firms. In order to continue their work on defense technology projects, these firms required either direct investment from the government, or a guarantee that the government would compensate them for their internal research.

The Contemporary Defense Industrial Base and the Revolution in Military Affairs

The revolution in military affairs (RMA) literature, discussed in chapter 1, set forth not only a general theory of military innovation—that military technology and doctrine exist in a kind of punctuated equilibrium, with long periods of relative stasis following short periods of intense change—but a specific theory of what the *next* revolution in military affairs would look like. The transformations envisioned by RMA theorists would have huge implications for the structure of the defense industrial bases of the United States, the Soviet Union, and beyond.

The events of the 1973 Yom Kippur War sparked much early RMA theorizing. In that conflict, Arab and Israeli forces used precision-guided munitions to devastating effect. In particular, the Egyptian use of surface-to-air missiles and guided antitank munitions nullified traditional Israeli advantages in marksmanship and armored warfare. On both sides, the effectiveness of weapons increased dramatically, making the conflict far more lethal than either side had expected. Although the war ended with a traditional armored offensive, the course of the conflict created serious questions about how armies and air forces would conduct conventional warfare in the future.[10]

These developments fell into line with a body of theory in the Soviet military circles about the emergence of the "reconnaissance strike complex."[11] Soviet theorists suggested that future warfare would depend on a combination of long-range precision munitions, advanced sensors and reconnaissance capabilities, and real-time command and control that could enable near-direct communication between "seers" and "shooters."[12] Such warfare might involve combat between competing reconnaissance strike complexes across wide fronts, each striking deep into the staging areas of the other. The destruction, at least in terms of reducing the organizational capacity of an enemy force, would approach that previously estimated to require the use of tactical nuclear weapons.[13]

This theory of the next RMA had the potential to have big effects on the military-industrial complexes of both the West and the Soviet bloc. On the doctrinal side, these developments helped revive the "operational" level of war, as opposed to the tactical and strategic. While the tactical level focuses on the means with which individual units fight one another, and the strategic level deals with the conduct of large-scale campaigns, the operational level concentrates on how organizations plan for and maintain themselves during extended engagements. In the United States, the operational manifestation of these theories became known as effects-based operations (EBO). EBO expected precision-guided munitions (PGM) attacks along the entire depth of enemy formations conducting either offensive or defensive operations.[14] The radical increase in information processing and communications technology, combined with the expansion of intelligence collection and aggregation capabilities into the upper atmosphere, would render the battle space more intelligible and more plastic than ever before, making the identification of critical targets possible.[15]

Increasing the precision of weapons and the capacity of sensors, and the ability of communications technology to transfer large amounts of data in a very short period of time, would require frequent escalations in processing speed.[16] It became clear that to make that possible, the military would need to be able to consistently rely on the rapid improvements being attained by the civilian tech industry. While the United States had long appreciated the implications of dual-use technologies for the relationship between national prosperity and military strength, in the 1950s and 1960s the contribution had mostly gone the other way; investments and advances in military technology could "spin off" into products with civilian application. Possibly most well known to the interested public may be the example of ENIAC. Construction on the Electronic Numerical Integrator and Calculator, which was invented at the University of Pennsylvania to calculate artillery firing tables for the US Army's Ballistic Research Laboratory, began during World War II.[17] The researchers applied for a patent on the ENIAC in 1947, and shortly thereafter formed the Eckart and Mauchly Computer Corporation. As Martin Weik writes, "ENIAC was the prototype from which most other modern computers evolved."[18] That company was subsequently sold to Remington Rand, major manufacturer of business machines and office equipment. And the military had a hand in the development of many technologies associated with the computer industry.[19]

But by the 1970s, the evident usefulness of civilian-led computing technology for military benefit led many analysts to conclude that this the flow of relationship was reversing itself more generally; primarily civilian-oriented technologies would provide the foundation of the West's military advantage.[20] The flowering speed and variety of 1970s and '80s hardware and software innovation was predominantly the germination of the seeds within industries not traditionally associated with the military-industrial complex.[21] Soviet military authorities soon concluded that Soviet industry lacked the ability to compete with the West on the key technologies of the

RMA, an appreciation that had some effect on the Soviet decision to reduce tensions with the United States in the mid-1980s.[22] The USSR's resultant collapse helped prevent a full test of these RMA concepts.[23] It was the 1991 Gulf War that fully demonstrated the promise of the long-range reconnaissance strike complex, as US and coalition forces wreaked extraordinary destruction on fortified Iraqi forces and governmental infrastructure with long-range air strikes and cruise missile attacks.

The success of coalition efforts appears to have confirmed RMA theories.[24] However, it was also apparent that, government support notwithstanding, the American NIS was insufficiently nimble to keep pace with the latest development in civilian computing technology.[25] This led some to argue that the DIB in its extant configuration could not even survive, much less dominate, the latest revolution in military affairs; faster-evolving civilian tech firms were seen to be replacing companies that concentrated on military contracting.[26] Secretary of Defense Les Aspin and other senior officials within DoD reasoned that the US defense industry of the late Cold War was far too large to sustain on a shrinking post–Cold War defense budget.[27] In a 1993 speech later dubbed "the Last Supper," Deputy US Secretary of Defense William Perry suggested to a group of defense industrialists that the number of major firms should shrink by about half, from fifteen to seven or eight. This talk, along with associated changes in Defense Department policy, helped fundamentally change the nature of the US military-industrial complex.[28] The next five years saw several major mergers and acquisitions between and among long-time major competitors: Rockwell and McDonnell Douglas joined Boeing; Martin Marietta joined Lockheed to become Lockheed Martin; Northrop purchased Grumman and became Northrop Grumman. The mergers led to a reduction of production facilities and infrastructure, but crucially allowed firms to capture the intellectual property and tacit knowledge owned by their competitors.[29] Yet, even with these mergers, traditional providers could not supply advanced computing and communications technologies at a rate and price competitive with commercial producers.[30] Moreover, additional legal problems with the DIB's technological synergies arose: pressure from the Department of Justice, concerned about antitrust issues and a lack of competition, forced DoD to reverse this policy in 1998.[31]

Pentagon studies since the turn of the century have captured the expected shift in the origins of innovations.[32] They specifically called for the Defense Department to expand access to civilian-sector innovation, both for those specific technologies and to spur innovation from the DIB.[33] As a result, the Department of Defense has made it a priority to transform the defense industrial base by reaching out to smaller inventive entities.[34]

Defense Firm as Aggregator

Procuring advanced technology from smaller civilian companies was inevitable. The increased power of the smaller firms is the result of the several advantages they have over large firms in pursuit of innovative technologies. James Hasik argues that uncertainty and a low-cost research and development can favor small firms over large. Fields like computing do not require huge cash investments for laboratories and other expensive R&D equipment. Moreover, small firms have more incentive to engage in risky long-range planning than do their large, established cousins, in order to generate market share in a crowded field.[35] Finally, smaller firms can be more competitive than their much larger cousins in fields where success is possible with a small, highly skilled workforce and in environments that favor flexible organizational decision-making.[36]

But while the logic of the RMA requires cutting-edge technologies that are the fruit of successful civilian firms, it also requires an active pursuit of systems integration.[37] The networked battlefield—which links sensors to stand-off precision-guided weapons—demands that all elements of the system speak the same language so that every weapon can talk to every other weapon. The military expectation has become that the role of the large defense firms will be as an aggregator of the products of smaller firms, each of which specializes in the production of components for the systems big firms eventually produce.[38] Rather than conduct the basic research necessary for the operation of the reconnaissance strike complex, the defense giants coordinate and integrate technologies into platforms (specific classes of ships, planes, and ground vehicles) and weapons.[39] Thus, dozens of firms contribute to the construction of a Lockheed Martin F-22 jet fighter, for example, most through subcontracting arrangements with the mother firm.[40] The rise of the defense technology firm as aggregator, and especially the acceleration of the aggregation trend in the 1990s, seems incompatible with the desire of the Pentagon to pursue relationships with small, nontraditional defense providers. However, expecting each aggregator to manage integration within their product reduces the total number of systems for the military to integrate.[41]

This trend toward defense contractor as aggregator has shifted the relationship of small firms with the traditional firms from that of relatively weak subcontractor to allied partner.[42] Any relationship between multiple companies in a high-technology field requires agreement on the initial ownership of intellectual property, and then on the owner's ability to control use of that property. But while the smaller partners are focused on creating and protecting their trade secrets and patent rights, the major aggregators want access to as much data as possible in order to ensure long-term viability of its contracted system. We discuss below the history and current state of ownership of defense technologies. Then we note that the viability of these business negotiations becomes even more fraught in the world in which the state secrets privilege hangs overhead.

Intellectual Property in the Modern Defense Industrial Base

The extent of an inventor's patent rights depends on the amount of government funding involved, and on which agency provides the funding.[43] Prior to 1980, the government typically took title to inventions that arose during performance of a government contract. However, as civilian technology played a greater role, there was a greater concern that eliminating an entity's ability to commercialize its invention was too great a disincentive to the development of important new technology. This realization led to the Bayh-Dole Act of 1980, which defaulted to a presumption that the government's rights to patentable technology developed for an invention conceived or built in performance of a government contract is most often limited to an irrevocable but crucially nonexclusive license. However, if a defense company contracts with the federal government, the contractor must adhere to the Federal Acquisition Regulation (FAR) to protect its patent rights.[44] Patent clauses affect the patent rights in all prime contractors and subcontractors under a government contract. Again, FAR clauses are "flowed down" to all subcontractors, regardless of tier. The flow-down allocates rights and obligations between the subcontractor and the government, not the subcontractor and the prime, unless the transaction is undertaken with commercial terms.

Nonetheless, the Bayh-Dole Act does give public and private actors broad discretion to negotiate over ownership and employment of intellectual property.[45] And some have argued that many of the problems that have emerged with respect to military contracting are the result of unrealistic negotiating positions rather than inevitable legal and business conflict; procurement managers could simply behave in less assertive and predatory ways, and thus ensure better relations with private suppliers.[46] Industry has regularly complained of overzealous policies designed to force contractors to give up their hard-won technical data.[47] One individual interviewed for this book expressed frustration at recurring conflicts between government and private enterprise, suggesting that both sides could share sufficiently large pieces of the pie.[48]

In 2018, Dr. Bruce D. Jette, assistant secretary of the Army for acquisition, logistics, and technology, established new guidelines for Army acquisition of intellectual property. Calling for a "balanced approach," Jette noted the difficulties that existing contracting procedure created for both the Army and industry, especially nontraditional providers. Jette established four principles: fostering open communication with industry, planning IP acquisition strategy based on the nature and availability of the equipment, preparing to negotiate flexible licensing agreements that preserve rights for both contractors and the Army, and setting prices as early as possible in the process.[49] Jette also argued that potential contractors should protect themselves during the negotiations process by holding their intellectual property as tightly as possible.[50] Although it is too early to determine the effectiveness of these reforms, they do indicate that the Army understands the value of intellectual property, and the problems outlined in this study.

Invention Secrecy

As discussed in chapter 2, states remain reluctant to relinquish their power to formulate patent policies to benefit state security interests. Governments are willing to reduce patent protection to the point of gutting it completely, when doing so serves national security interests. This section discusses how the US government specifically has obstructed patent protection and, in some cases, has entirely eliminated the patent holders' ability to enforce their patents against unauthorized use. It focuses on the Invention Secrecy Act of 1951, as well as the state secrecy privilege.

Patent protection drove the innovation of many key technologies in World War I, including the flamethrower (patented both in Germany and the United Kingdom), the Brodie helmet, and the Lewis machine gun.[51] Unsurprisingly, states displayed a willingness to ignore foreign claims of infringement regarding these technologies.[52] For example, during World War I, the US Navy sought to have its military contractors build a wireless radio system. There were US companies capable of manufacturing that technology, but they hesitated to do so because of potential threats from the patents of the Marconi Corporation, a company located within US ally Italy. Congress responded to that hurdle by immunizing the domestic companies from any related patent infringement lawsuits.[53]

Given how willing the United States was to discard the patent rights of allied patent holders, it is unsurprising that it was willing to do so for enemy *foreign* patent holders as well. The United States seized a variety of German patents at the advent of its participation in World War I, a process it would repeat when it entered World War II.[54] More surprising was the decision to permit the nullification of potential rights of even *domestic* inventors. Yet in 1917 Congress chose to do so, authorizing the US Patent and Trademark

Office to classify certain patents beneficial to the war effort as "secret."[55] Upon issuance of a secrecy order, the government froze issuance of a patent, prohibited the filing of foreign patent applications, and ordered the applicant inventor to keep the invention secret.[56] The patent law security provisions threatened an inventor who sought to obtain foreign patent rights in light of such a secrecy order with criminal punishments and the loss of future US patent rights.[57] Of course, the statute's drafters recognized that such secrecy orders might cause inventors to turn to a trade secret approach rather than seek patents.[58] So, to diminish the negative impact of a secrecy order, the 1917 Act contained a provision requiring the government to provide compensation if a patent was later granted *and* if it was acknowledged that the *government* used the invention. The inventor received no compensation if the patent never issued because suppression continued indefinitely, or for any private use of the invention unless contracted for by the government use.[59] While Congress ended these secrecy provisions shortly after the end of World War I, President Franklin Roosevelt reestablished them just prior to the US entry into World War II.[60] During the World War II, the number of secrecy orders in effect peaked at 8,293 on December 31, 1944.[61] At the end of the war, as the likely extended duration of the Cold War became clear, the US government recognized the need to have similar secrecy powers going forward.

Much of this emphasis on secrecy was motivated by the desire to protect nuclear weapons technology, as post–World War II nuclear technology was treated very differently from the other technology that had been suppressed during the war. Recognizing that many of the inventions had dual-use potential, the secretary of commerce explained to Congress that continuing the secrecy orders unnecessarily impaired the commercial use of those inventions. Accordingly, the commissioner of patents rescinded most of the existing patent secrecy orders, approximately 6,575 in number.[62] By contrast, the government explicitly rejected disclosure of nuclear technology. In August 1945, shortly after the US Army Air Force dropped the first atomic bomb on Japan, Secretary of War Henry Stimson released a statement praising the efforts of scientists and military officials that led to the bomb's successful development. Stimson also noted that the government had already concluded that the patent system provided potentially dangerous access to America's nuclear technology.[63] Unlike the released majority, applications directed to nuclear technology, approximately eight hundred in number, remained under secrecy orders.

But just a few years later, the government's view shifted considerably; it apparently concluded that many forms of technology were strategically important. With passage of the Invention Secrecy Act of 1951 (ISA), the government's authority to maintain the secrecy of *any* inventions implicated in national security was made permanent.[64] The ISA outlines three differing legal statuses—wartime provisions, national emergency provisions, and peacetime provisions—and differing secrecy order provisions for each.[65] During a war, the ISA authorizes an essentially unchallengeable order to "remain in effect for the duration of hostilities and one year following cessation of hostilities."[66] As formal declarations of war tend to be limited in time, it was a conceptually powerful but cabined authority—and, as Congress has not formally declared war since World War II, this section has been impotent. In 1950, however, President Truman declared a national emergency that would last for the next twenty-nine years.[67] As a result of this declaration, secrecy orders increased dramatically. In 1950, approximately 2,395 patent applications were under secrecy orders.[68] From 1951 to 1958 the number of secrecy orders rose from 3,435 to 6,149, and it remained between 4,100 and 5,100 for the next twenty years.[69]

Ostensibly, the ISA established two alternative standards that the government must satisfy in order to authorize suppression of a patent, depending on whether the government has a property interest in the private invention—generally, where the government has worked with private industry. If that is the case, an agency must at least show that publication or disclosure of the invention "*might* . . . be detrimental to the national security."[70] Although this standard theoretically provides a floor below which suppression could not be justified, it is so porous that the government's discretion is effectively unfettered. By contrast, the statute formally establishes a higher standard for "John Doe" suppression orders where the government does not have property interest because the inventors' work has been wholly independent of the government. An agency must show that publication or disclosure "*would* be detrimental to the national security"—a nominally more stringent standard than the conditional "might be." While the government's frequent involvement in military technological development makes the formal relevance of this standard slightly less common than the former, it is hardly rare; since 1988, more than 41 percent of all secrecy orders filed have been John Doe orders.[71] In practice, however, the standard is the same—which is to say that the government's authority to invoke a secrecy order goes effectively unchecked.

An individual patent applicant can request a secrecy order rescission from the defense agencies that issued it, and such a request will trigger a review of the order.[72] However, such a review is rarely successful. The secrecy order review process normally happens within the first few months after a patent application has been filed.[73] Unsurprisingly, the people determining the appropriateness of a secrecy order are involved in the defense agencies' research and development, as those are the employees with sufficient knowledge to properly evaluate the national security value of any new patent applications.[74] But one can expect these same defense officials to be tempted to pass along the information gained during a secrecy order review to their established contractors for use in proposed or modifiable defense technology, particularly if such officials can look forward to employment with those contractors after they retire from the public sector.[75] Yet the courts have rejected the few attempts to dispute the circumstances in which the government may conclude that an application is subject to being withheld as potentially endangering national security. This is hardly surprising; as indicated by the *Lucent v. Crater* case, the courts traditionally give the executive branch very wide discretion on issues of national security.[76]

Only in 1979, when the National Emergencies Act became effective, did the ISA's peacetime provisions take effect for the first time in ISA's almost thirty-year existence. Under peacetime, the statute requires annual review of each patent to ascertain whether continuation of the secrecy order is in the national interest. The result of this policy shift was a cautious reduction in intellectual property classified as secret: the first review, undertaken between September 1978 and March 1979, resulted in 1,150 declassifications and 3,300 renewals.[77]

The reform did not settle the issue, however. During 1979, the Patent Office received 107,409 patent applications. It concluded that 4,829 of those applications might have been of interest to defense agencies, and it sent them to DoD for review. Two hundred forty-three secrecy orders were issued, 200 of which contained security classification markings when filed.[78] Such significant use of secrecy orders in the absence of a formal national emergency caused the US House of Representatives to hold hearings on their appropriateness in 1980, during which it was noted that the original enacting body "never set down a rationale for invention secrecy."[79] Much of Congress's concern appeared to be related to the danger of impeding civilian economic development—perhaps unsurprising, as that was around the time that the United States first found itself subjected to increasing technological competition from Japan. But the executive branch denied that there was significant civilian value to the intellectual property being suppressed, noting that most of the suppressed inventions were created in conjunction with government contracts, generally for defense purposes.[80]

In theory, whether the suppressed inventions have civilian application would not adjust the calculus regarding potential injury to the inventors, because the ISA, like earlier wartime statutes, authorizes inventors to receive compensation for lost profits from government use caused by the order of secrecy. However, perhaps in recognition by the government of the importance of dual-use technology, it also permits recovery from the government for lost profits from *private* use. Moreover, unlike in the previous acts, inventors do not have to have first offered their inventions to the government in order to assert claims relating to private use.[81] But while this might suggest a sufficient compensation scheme, it has come under fire from several angles. As noted by the relevant courts hearing such cases, inventors must prove actual damages.[82] Satisfying this standard requires proving both that the unauthorized user would not have been able to substitute an alternative sufficient to effectively compete, and proving how much the inventor would have reasonably obtained from licensing and/or competitive sales.

In the late 1980s, a major generator of secrecy orders was shut down. Interviews conducted by Dorothy McAllen of two government workers employed in the aerospace/defense sector suggest that around that time, their agency issued an internal policy ceasing the submission of patent application unless there was clear future commercial value to the invention. The government was always more inclined to issue a secrecy order for any technology that it alone owned, as there was no private inventor to demand issuance of the patent or compensation. This decision, rather than any major shift in bureaucratic philosophy regarding secrecy orders on truly civilian-born dual-use inventions, is the cause of the decline. In fact, individual inventors face the same roadblocks to getting their IP patented and compensated in the post–Cold War period as they did during the Cold War.[83]

However, while the total number of secrecy orders filed each year dramatically decreased from 630 in 1988 to 86 in 2010, it does not appear that the government has lost interest in seeking secrecy orders regarding wholly privately developed inventions.[84] There were 5,680 invention secrecy orders in effect at the end of fiscal year 2016, with 121 new secrecy orders issued in 2016.[85] Recent direct attack on the act itself has so far been unfruitful.[86]

Contracting Practice

The structure and practice of intellectual property law can have an impact on the structure of the defense industry by affecting the behavior of private firms. The government wants small technology firms to focus their innovative efforts with defense applications in mind, and it holds out the promise of substantial government sales as motivation.[87] Private firms have several reasons to worry about their intellectual property, and to resort to the use of IP law to protect themselves from other firms (and from the government).

If a private company wants to sell to the civilian market as well as fulfill government contracts, it needs to ensure its continued control over its own trade secrets, lest competitors gain access to the data and processes needed to replicate the company's products. Private firms operating in partnership with defense firms also need assurance that their IP will remain safe from their partner. This requires a confidence in a judicial system that can enforce those rights if disputes arise. On the other hand, some have argued that small civilian-oriented firms have displayed considerably more sophistication with respect to IP contracting than their large defense counterparts.[88] So why haven't small, flexible, nimble private firms replaced the lumbering dinosaurs of the defense industrial base? Over the past two decades, the Department of Defense has made repeated efforts to encourage small firms to engage with DoD, and to improve the standing of small firms with respect to their larger competitors.[89] Donald Rumsfeld reiterated this case in the 2000s, as did Secretary of Defense Ash Carter during the Obama administration.[90]

The Office of Small Business Programs (OSBP) is one manifestation of this interest. The OSBP was created by statute in 1978 as the Office of Small and Disadvantaged Business Utilization, and took its current name in 2006.[91] According to its mission statement, OSBP advises the secretary of defense on matters related to small business, and attempts to maximize the contributions of small business to DoD. Organizationally, OSBP is within the Office of the Undersecretary of Defense for Acquisition, Technology, and Logistics. The office seeks to develop small businesses as suppliers to DoD (and to front-line war fighters) and to increase the percentage of DoD funding that goes to small business.[92]

In an April 2015 speech at Stanford University, Secretary Carter elaborated on the Defense Department's view of high technology contracting.[93] Carter made an explicit point about the role of intellectual property in the relationship between Silicon Valley and the Department of Defense. He tried to allay the fears of entrepreneurs:

> One concern I've heard about is the worry that the government will insist on taking intellectual property, and then reveal proprietary information to the public and to competitors. Let me assure you that we understand and appreciate industry's right to intellectual property. And DoD has a long history of successfully protecting companies' proprietary information, and we respect the fact that IP is often the most important and valuable asset a company holds, and that businesses cannot be forced to sell their IP to the government. We understand all that. We need the creativity and

innovation that comes from start-ups and small businesses, and we know that part of doing business with them involves protecting their intellectual property.[94]

The foundations of innovation now lay, Carter argued, primarily in civilian hands. At the same time, government has played an important role in founding the digital world. To facilitate cooperation between Silicon Valley and the Department of Defense, Carter has activated a Defense Innovation Unit-Experimental (DIUx) in Mountain View, California. Led by a senior employee from Defense Advanced Projects Research Agency (DARPA) and a former Navy SEAL, the DIUx hopes to serve two purposes: identify advanced technology, and ameliorate the technology industry's concerns about the Pentagon.[95] George Duchak, the first director of DIUx, referred to the unit as "match.com" for the Defense Department, with a focus on bringing nontraditional providers into DoD's orbit.[96]

Unfortunately these reforms seem to have consistently failed, as Pentagon procurement remains dominated by the small number of large, traditional defense firms. These firms continue to enjoy advantages for several reasons.[97] First, the traditional defense providers have mastered the complex rules that govern defense procurement. In order to ensure fairness and prevent corruption, Congress and the executive branch have established an extremely detailed system for managing every step of the process, and companies that have no familiarity with this process are at a huge disadvantage against those that do. Moreover, even in a noncompetitive situation, the prospect of dealing with the Pentagon's procurement system may prove daunting to nontraditional providers.[98]

Second, while modern military equipment increasingly involves the adoption of technologies with civilian as well as military applications, or "dual-use" technologies, the tolerances for military equipment generally go well beyond those required for civilian products. While civilian and military products often seem similar, the military versions generally need to survive in much more difficult circumstances, and perform more complex tasks. Traditional defense providers understand these requirements and design equipment with them in mind. Civilian firms, on the other hand, have more trouble adapting their products, and this puts them at a disadvantage against traditional defense firms, whose products will require fewer adjustments to military use.[99]

Finally, traditional defense firms tend to employ large numbers of personnel with experience either in the military, the procurement office of the Department of Defense, or the congressional staffs tasked with evaluating weapons systems for federal funding in what is commonly referred to as the "revolving door." And this door is spinning faster. According to a 2010 study in the *Boston Globe*, "from 2004 through 2008, 80 percent of retiring three- and four-star officers went to work as consultants or defense executives, according to the Globe analysis. That compares with less than 50 percent who followed that path a decade earlier, from 1994 to 1998."[100] On the upside, these relationships help the DIB understand DoD needs, and help the DoD understand industry limitations. Admirals, generals, and procurement officers, and senior staffers have sufficient professional and subject-matter expertise to make valuable contributions to their employers' understanding of the procurement and development processes. These contributions give traditional defense providers an advantage in both the innovation process (producing better weapons) and the competitive process (ensuring that DoD selects a firm's systems).

Even without the revolving door, government officials are generally interested in maintaining good relations with these regular suppliers. As a result, even when one of these large companies holds the intellectual property rights to an invention useful for a military system and is not the party contracted to produce that system, the military will likely agree to license that invention rather than try to undercut one of its regular suppliers. In contrast, smaller businesses do not have the political and economic influence to compel the government to license their intellectual property. Instead, they are left feeling impotent and apprehensive. Indeed, one commenter on the respected blog Patently-O described this impotence through the following apocryphal dialogue:

> Patentee sends letter to [the government] contractor saying "You are infringing my patent." [*sic*]
> Contractor sends letter to patentee saying "Pound sand, weasel. Go take it up with the [the government,] 'cuz it ain't my problem."
> Patentee send letter to [the government] saying "My patent is being infringed by the work being done under your contract with the contractor."
> [The government] replies "So sue me. I will consign your claim to a 'black hole' where it will languish for years as an administrative claim."
> Patentee send letter saying "Here is my claim."
> [The government] replies "Got it. We will get back to you sometime in the next decade of [*sic*] so. Hopefully, by then you will be under Chapter 11."[101]

To the extent that this view is prevalent among small businesses, they will anticipate the need to rely on legal recourse rather than clout. However, innovators attentive to that prospect will be dissuaded by the possibility that, if the fruits of their efforts are used without authorization, their avenues of recourse could be blocked through the government's use of the state secrets privilege. As described earlier in the chapter, this privilege is a common-law (as opposed to statutory) creation that allows the government to refuse to disclose possible use of technology during intellectual property litigation that would otherwise be required, as long as the government merely asserts that there is a reasonable danger that such disclosure would harm the national security of the United States.

The privilege emanates from a series of nineteenth-century judicial rulings with an underpinning rationale that there is a public interest in allowing the government to shield its military and diplomatic strategy from its enemies.[102] But its establishment was not cemented in the area of military technology until the case of *Reynolds v. United States* in 1953. In this case, three employees of an Air Force contractor were killed when a B-29 Superfortress crashed. The employees' widows sued the government under an act that would permit them to recover—but only if the Air Force had engaged in negligence. To prove negligence, the widows had sought to obtain from the Air Force the official post-incident report and survivors' statements. The Air Force opposed disclosure of those documents, arguing that the aircraft and its occupants were engaged in testing new radar equipment, and that release of those reports to the widows' counsel would threaten national security. The government would not even provide the reports to the district court to permit the judge to evaluate the reasonableness of the government's assertion. The government's refusal led the court to make a presumption of negligence against the government and to direct a verdict in favor of the widows. Even after that decision was affirmed by the circuit court, the government steadfastly maintained its refusal to produce the document.

The Supreme Court upheld the government's refusal, declining to order that the government produce the withheld documents, and precluding a presumption of negligence from the failure to do so.[103] Just fifty years later, the accident report's release shed a very bright but harsh light on the government's use of the privilege. That report disclosed exactly what the plaintiffs had alleged: that the crash had indeed been caused by the government's negligent maintenance of the plane. Moreover—and more significantly, in evaluating the government's motives—wholly contrary to the government's position, the accident report contained no information concerning the allegedly confidential radar technology. And the only reason that this had not become known during the case was because the Supreme Court had refused to compel production of the report, even to the justices. Essentially, it become clear that the entire foundation of the modern state secrets privilege was built on a lie—one about which the *Reynolds* court was, if not complicit, willing to be snookered.[104] And it demonstrated that the government was perfectly willing to employ that privilege to accomplish goals other than protecting national security.

State Secrets Privilege

As the state secrets privilege has developed, both the circumstances of its invocation and its impact on litigation have had far broader effects than originally expected.[105] Indeed, invocation of the privilege has increased in recent decades. From 1953 to 1976, it was invoked only *four* times, but in the five years from 2001 to 2005 it was invoked *twenty-three* times. In the Senate Judiciary Committee's report on the proposed State Secrets Protection Act, the committee noted that "the Bush administration has raised the Privilege in over 25 percent more cases per year than previous administrations, and has sought dismissal in over 90 percent more cases."[106]

Employment of this privilege has been fostered by the lack of judicial oversight. While the Supreme Court did declare that an invoking agency head would be required to provide the deciding court with information surrounding the creation of the document, it declined to require that the executive branch provide the courts with the underlying information as proof of the danger to security. Thus, while a judge may request that the agency produce the information for her review, the agency is free to refuse. The *Reynolds* court contended that "too much judicial inquiry into the claim of privilege would force disclosure of the thing the privilege was meant to protect," though it is difficult to imagine the circumstances in which permitting review by a district court judge—an Article III appointment, vetted through prior executive branch review—is likely to be equivalent to the type of "disclosure" that was originally of concern.[107] In practice, the government can always provide a seemingly plausible reason for invocation of the privilege, a reason that can withstand superficial evaluation. And while courts have indicated that a private litigant's need for the information may be relevant to the amount of deference afforded to the government, once a court concludes that the privilege has been properly invoked, "even the most compelling necessity cannot overcome the claim of privilege if the court is ultimately satisfied" that the privilege is appropriate.[108]

This litigation tactic was developed contemporaneously with the development of the Invention Secrecy Act, and although no evidence indicates more than a coincidental connection, the state secrets privilege has become a key method for the government to avoid having to defend against patent infringement lawsuits. The need to rely on the government's assertions absent the opportunity to evaluate the underlying facts created a rubber-stamping process particularly in patent cases. The *Reynolds* court noted that the government's ability to provide equivalent information to the plaintiffs while avoiding disclosure is a key factor in determining whether the agency's invocation should be upheld, but even evaluating the equivalence of the information is in practice dependent upon the government's conclusions as to equivalence.[109] Thus far, in patent litigations a court has rarely, if ever, required production of the documents to the private patent holder even when there has been no other way to confirm or deny the government contractor's use of the patented device or method.[110]

This result leaves the patent holder attempting to litigate a claim of infringement against the government with one hand tied behind their back. And in practice, in many contract cases the government has effectively turned the privilege into full handcuffs: an all-purpose get-out-of-infringement-cases-free card. By claiming that it could not effectively mount its defense unless it used (and thus disclosed) the protected information, the government has regularly convinced courts to dismiss cases even where the plaintiffs had been willing to go forward without the disclosure.[111] In 2011, for example, the federal government asserted the privilege to prevent the disclosure of sensitive stealth technology in a defense contract dispute with a government contractor.[112] In refusing to find an enforceable contract "where liability depends upon the validity of a plausible . . . defense, and when full litigation of that defense 'would inevitably lead to the disclosure of' state secrets, neither party can obtain judicial relief." The nail in the intellectual property owners' coffin is the government's recent argument that the state secrets privilege in this case is not limited to certain specific documents that contain confidential information, but rather covers the information (namely, the government's choice of technology) itself.[113] In nearly all intellectual property disputes, the primary issue is the government's choice of its technology.[114] As a result, these arguments prevent successful pursuit of most claims of unauthorized use by intellectual property holders, regardless of whether the patent holders might have access to other sources of information that might reasonably lead them to succeed. In about 73 percent of patent cases the court has upheld the assertion in full, and in another 10 percent it has upheld it in part.[115]

Therein lies the problem: the government's use of the privilege to foster uncertainty in a commercial licensing context is dangerously counterproductive. Some of those on the federal court most often hearing these cases have recognized the harm being done. The esteemed Judge Newman, of the US Circuit Court for the Federal Circuit, who authored a dissent to the Federal Circuit's *Crater* decision, argued: "Fair resolution of disputes is necessary to ensure the government's continued access to the private sector's talents.' Unexpected termination of disputes through the privilege undermines such a fair resolution.

However, it is even more counterproductive than perhaps even Judge Newman realized, because the smaller entities on which the government intends to rely lack the political clout and long-term institutional relationships to ensure that the military will properly license their technology. As a result, unless they can rely on having their intellectual property protected, small inventors will likely be disinclined to pursue inventions sought after by the military, or to make their nonpatented civilian inventions available to the military. Thus, injudicious application of the privilege discourages innovation by those whom the Defense Department most wishes to encourage.

In sum, this legal concept has become a means through which the government and major firms within the defense industrial base protect themselves from the claims, particularly claims by the "Davids"—small, innovative, civilian-oriented firms like Crater. Oddly, use of the privilege continues, and indeed has expanded, even as the Department of Defense has called for small firms that have not traditionally participated in defense contracting to contribute more to the US national innovation system. The result, however, is that this privilege places hard constraints on government endeavors to make its defense industrial base more innovative.

Conclusion

Intellectual property protection is embedded at every level of the US defense industrial base and national innovation system. The constellation of private companies, research universities, and government labs that constitute the manner in which the United States develops and builds weapons requires close attention to the ownership and transfer of technology. This system, internationally idiosyncratic for a variety of reasons, excels at certain tasks while lagging at others. Its drawbacks have driven calls for reform from a sequence of secretaries of defense and other major figures, but successful reform threatens to break certain aspects of the system.

4: INTELLECTUAL PROPERTY IN DEFENSE IN COMPARATIVE CONTEXT

Vignette: Tony Stark and Borrowed Technology

Tony Stark, a character in the 2010 Hollywood film *Iron Man 2*, is the quintessential fantasy figure of American capitalism: a billionaire member in good standing of the military-industrial complex, whose father, allegedly single-handedly, created the underlying technology behind the Iron Man suit, Stark's signature weapon.[1] Yet in the film, that inherent American superiority is thrown into question by an allegation that the technology was truly created in conjunction with an unsung Soviet scientist named Anton Vanko, thus giving Vanko's son and heir Ivan some claim to the intellectual property within the suit.[2]

Regrettably, Stark does not place his faith in the legal system in order to defend his claim against Vanko.[3] And, while trying to fend off Vanko's claim through other means, Stark is pressured by the US government to give up the secrets of the Iron Man suit. After Stark refuses a senator's demand that he relinquish his body-armor technology, he risks billion-dollar pieces of equipment to impress guests at a birthday party. The US government uses that reckless debauchery as a pretext to assert full control of the equipment, only to turn it over to a competitor that then uses the technology to fulfill its own defense contract.[4]

This fictional world gives a surprisingly accurate view of some of the issues that have faced and are facing inventors in the United States and abroad. Though Stark quickly dismisses Vanko's claim, many of Russia's best inventors did indeed face major challenges in reaping the benefits of their work due to the Soviet Union's obvious lack of individual intellectual property protections. Mikhail Kalashnikov "invented" the AK-47, but never profited directly from the sale of nearly one hundred million weapons. Similarly, Alexey Pazhitnov created the video game Tetris, only to watch as the Soviet government sold the rights to Nintendo. Although both Kalashnikov and Pazhitnov managed to prosper in the end, the idea of the brilliant Soviet inventor who watches his invention used by others and receives no compensation for it is hardly outlandish.[5]

Similarly, in the United States the path from defense inventorship to profits does not run straight. In real life, most inventors aren't trying to "privatize peace." Many just want to get a government contract. Inventors are supposed to profit from their creations. Their inventions are kept as trade secrets, like the Iron Man suit, or are patented. But, consciously or no, *Iron Man 2* echoed the real world; the US government can take such actions with almost total legal impunity. It is not unheard of for a potential contractor to provide the government with product specifications, only to then watch it award the contract to a competitor that has suddenly and suspiciously generated remarkably similar technology. For example, during World War I the British Royal Navy turned over patents on torpedo technology invented by Sydney Upton Hardcastle to British allies, precluding Hardcastle from patenting his work in France or the United States.[6]

As discussed in chapter 3, when invoking the military and state secrets privilege, the government claims that access to information about the military's use of technology might endanger national security. So in *Iron Man 2*, Stark's competitor, Hammer, can immediately begin production after acquiring the Iron Man suit under a government contract. If Stark sues, the government could claim state secrets privilege, protecting details of the contract and production design from Stark's lawyers. Of course, when the developer of such trade secrets is a well-connected defense contractor like Stark Industries, institutional relationships with the Pentagon will tend to protect its interests—which is why those who suffer this type of abuse are not suave billionaires, but small companies or even individual inventors.

Stark doesn't need a monetary incentive to develop his technology, and apparently neither does Vanko. And of course, both Stark and Vanko are comic book characters. But movie audiences should keep in mind that the claims of real-life inventors who correspond to Vanko are not, from a historical point of view, absurd. Nor is the concern absurd that states might have legal tools for acquiring and distributing the technology they want.

Introduction

This chapter examines the role of intellectual property protection in three major military-industrial complexes: Russia (and the Soviet Union), China, and the Republic of Korea. The chapter begins by discussing some themes and problems common to all three systems, including question of state versus private ownership, and of the relationship between government defense bureaucracies and private entities. The chapter then focuses on a close examination of the Soviet/Russian, Chinese, and South Korean systems, in that order. The first two merit inclusion because of their intrinsic importance to the innovation and diffusion of military technology around the globe. The third represents an interesting case of a mid-tier arms producer and consumer that has undergone its own transition with respect to intellectual property protection. All three systems have evolved over time, with particularly striking transformations in the first two. Each of these countries has taken a different road to the development of its national innovation system (NIS), both in its government and in its defense industries.

State-Owned Enterprises and Private Enterprises

Every NIS has its strengths and weaknesses, just as every configuration of intellectual property protection has its costs and benefits. Similarly, every modern defense industrial base (DIB) requires a degree of state intervention. Because of the vagaries of the international arms trade and the dangers of the international system, many states lend considerable financial support to the defense industries. The response to these problems has manifested in a variety of different means of managing the relationship between state and private investment. Some states have tried to solve the public-private problem by simply taking full ownership of their defense industries, though, as we shall see, this decision still produces some complexity with respect to intellectual property protection. Most states have settled on systems of public-private collaboration, which, as discussed in chapter 2, inherently create intellectual property protection complications. Finally, as intimated in chapter 1, the increasing importance of dual-use technology to military affairs has made it necessary for states to develop means for taking military advantage of private-sector innovation.

In the twentieth century, several different types of states experimented with defense technology enterprises wholly or partially owned by the state. In the socialist sphere, state ownership became the norm; heavy industrial production came wholly under the ownership and management of the government. Such Chinese and Soviet Russian firms as Shenyang, Chengdu, Sukhoi, Tupolev, Mikoyan, and Gurevich are characteristic representatives of state-owned defense enterprises (SOEs). State-owned defense technology firms resolve most of the major intellectual property problems associated with technological innovation. The firm owns any intellectual property, and the state owns the firm. As discussed in chapter 1, the AK-47 assault rifle resulted from competition between several design teams within the same firm, with the teams liberally borrowing features from each other's designs. The individual engineers had no property rights to the rifle (a situation also common to Western-style private firms), and the firm had no independent rights against the government. Economic systems with state-owned enterprises also tend to dissuade small firms from competing with the larger state-run firms, especially for major military contracts.

Nevertheless, the details of the structure of intellectual property practice can affect the behavior of state-owned firms, and especially of subcomponents within those firms. State ownership means that individuals and units receive rewards different from those allocated by the private market. Instead of enjoying market-driven financial success, state-run firms compensate units and individuals may receive privileges, promotions, prestige, autonomy, and resources. Consequently, even within state-owned enterprises, units have an incentive to protect their own inventions. If one unit within a state-owned firm suspects that another unit within that firm will steal its inventions without offering appropriate credit or recompense, it becomes unwilling to share its innovations. The result can be a system-wide decline in innovation, which is why most national innovation systems based on state-owned enterprises nevertheless offer some degree of patent or other intellectual property protection for inventions, thus giving particular firms an incentive to innovate and to share their inventions.[7]

DIBs built around state ownership stand in contrast to the US system (described in chapter 3), which is built around private firms that compete for contracts from the government. These contracts can involve agreements to produce specific weapons, but also agreements on research funding for future weapon projects. The details of a system built around private industry vary considerably, but in all cases they involve substantial government regulation of contracting, production, and sale. In this system, the state and private industry share a responsibility for innovation. Because of the necessity of establishing property ownership for various public and private entities, these systems necessarily require some specification of intellectual property rights.

The Russian National Innovation System

The NIS developed by the Soviet Union in the ashes of the Bolshevik Revolution helped to produce a huge amount of remarkably sophisticated weaponry. Working from a small, backward base, it began to produce weapons competitive with those of the West by the early 1930s. While the USSR regularly imported technology from Western sources, the Soviet DIB developed the capacity to generate enormous quantities of advanced military equipment during World War II and the Cold War. Despite heavy state investment, however, the Soviet NIS could not keep pace with American innovation in electronics and computing toward the end of the Cold War. The

Russian NIS that survived the collapse of the Soviet Union continues to produce effective equipment at competitive prices, and has worked to integrate new technology, and a new legal framework, into existing practices and infrastructure.

The Classic Soviet National Innovation System

At the end of the Russian Revolution, the Soviet state nationalized arms production, along with most other heavy industry. Several major firms remained semi-independent from one another, but all were owned by the state, and all had close relationships with the Red Army. Through the interwar period, World War II, and the Cold War this system developed into a sprawling array of manufacturing complexes, labs, factories, and testing grounds, dominating large tracts of the Soviet landscape.[8] In World War II, war material constituted 60 percent of the Soviet economy.[9] This percentage remained high in the postwar period, and throughout the Cold War. Although estimates vary (and Soviet data is misleading), defense spending made up as much as 40 percent of the Soviet economy as late as the 1980s.[10] Soviet industry responded to the demands of both the civilian leadership and the Red Army leadership. The requirements of the Red Army, in particular, shaped the contours of the industry and gave it the political power to outcompete civilian priorities. Theoretical work in the Red Army produced the "pull" necessary for innovation, with theoretical developments leading to contracts and investment.[11]

Soviet industry produced an enormous volume of weapons, many of which compared favorably with Western systems. Even in the wake of the disruption of the Russian Civil War, the Soviet DIB developed and produced a wide array of systems, including T-26 light tanks, BT-7 medium tanks, Polikarpov I-5 biplane fighters, and I-16 monoplane fighters. Some of these were derivative of foreign designs and included licensed technology, imported before the development of strong export control in the West. Military analysts generally regarded these systems as competitive with those of foreign contemporaries.[12] On the eve of World War II, the Soviet DIB began to produce some of the most innovative weapons in the world, including the Ilyushin Il-2 ground attack aircraft, the KV-1 heavy tank, and the T-34 medium tank.[13]

Soviet productivity continued into the Cold War. The Soviets churned out ground combat vehicles, small arms, and aircraft in numbers considerably greater than their Western counterparts.[14] Although exact comparisons are difficult, they produced roughly double the number of fighter aircraft as the

United States during the Cold War. Similarly, from a relatively slow start, the Soviet shipbuilding industry took off during the Cold War, producing an array of vessels from patrol craft to submarines and oceangoing cruisers.[15]

The Soviet NIS proved adept at some kinds of innovation, but it often lagged behind the West and required regular injections of technology. In the prewar period, the USSR acquired technology through direct licit or illicit purchases of military equipment.[16] After the war began, it enjoyed transfers of technology and weapons from the West (in some cases stealing technology, such as the B-29 Superfortress).[17] During the Cold War, Soviet efforts at acquiring Western technology mostly took the form of industrial espionage, due in large part to the imposition of export controls by the United States.[18] These controls, to be discussed at more length in chapter 5, prevented the USSR from acquiring certain categories of US technology.

INTELLECTUAL PROPERTY IN THE CLASSIC
SOVIET NATIONAL INNOVATION SYSTEM

The Russian Revolution spurred several waves of reform of Imperial Russian patent law.[19] Existing patent law in 1917 hewed close to European standards, but the revolution and its attendant disruptions made enforcement nearly impossible. The first wave of reform gave the Soviet state the right to appropriate any patented inventions; this had the result of pushing inventors outside of the system of IP protection, as the state appeared more dangerous than any potential pirates. The New Economic Policy of 1924, needing to jump-start the moribund Soviet economy, restored most patent rights to individual inventors; but the end of NEP in 1931 curtailed some of these rights, and in any case the lack of private capital made it difficult for inventors to take advantage of their patents. However, the Soviet Union did join the Paris Convention for the Protection of Industrial Property in 1965.[20] Soviet IP law made virtually no allowance for protection of trade secrets, but Soviet inventors had access to the IP systems of other countries. Indeed, inventors would sometimes bypass the Soviet bureaucracy and apply for patents directly in the United States.[21]

By the early 1930s the USSR established its Committee for Inventing, which was responsible for overseeing all inventions, nonsecret and secret. In 1936 the Soviet government abolished the committee and established a bifurcated legal framework for Soviet inventions, placing innovations relating to the military and national security under the auspices of the military and certifying them as "highly classified" by the Ministry of Defense. A civilian authority handled all other inventions.[22]

Despite possessing what would seem to be the type of bureaucratic structure that would retard innovation, as judged by the applications for "highly classified" certification, the pace of development of Soviet military technology appears astounding. By 1967 more than ten thousand applications were filed requesting a "highly classified" certificate, and by 1987 the annual totals were close to thirty thousand.[23] These applications are estimated to have resulted in about seventeen to twenty thousand secret inventors' certificate grants annually toward the end of the 1980s. By contrast, the number of approved US secrecy orders regarding technology amounted to no more than a few hundred each year.[24]

What explains this difference? Engineering groups would likely have been the target of significant domestic political pressure to be seen as productive, especially for the military's benefit.[25] Because of this motivation to justify labeling any change as a breakthrough worthy of secrecy protection, the rate of alleged Soviet military innovations is more accurately described as incredible—as in the Latin *incredibilis* (not to be believed). This sole reliance on an invention secrecy system, rather than any use of a public intellectual property regime, likely affected diffusion of military technology *within* the USSR. Different parts of the Soviet military-industrial system struggled to benefit from innovation because they lacked access to the latest inventions and techniques *even within the Soviet Union*, much less the rest of the world. These restrictions almost certainly significantly hampered the evolution of those technologies, slowing down the pace of innovative improvements within the Soviet/Russian sphere.[26]

Moreover, by relying on secrecy classification, the Soviet Union and Russia faced the same problems and limitations faced by private actors who rely on its private equivalent, trade secrecy protection—namely, the inability to use legal systems to prevent foreign state actors who discover the technology from using it. This approach has meant that there have been no legal mechanisms to prevent diffusion of technology by those who acquire the technology through independent means. Indeed, unlike formalized trade secret protection, the Soviet Union left itself without access to a legal regime that might prevent some foreign use of the technology after improper appropriation, either through theft or through other unauthorized acquisition, such as illegal arms sales.

INDUSTRIAL ESPIONAGE AND THE SOVIET NATIONAL INNOVATION SYSTEM

In order to remedy a lack of internal technological innovation, the Soviet Union borrowed liberally from foreign systems. Chapter 6 of this book will detail one example of this borrowing: the B-29 Superfortress heavy bomber, which became the Soviet Tu-4. Another well-known example involves theft of secrets related to the US atomic bomb program, which helped accelerate the Soviet program.[27] The Soviets also purchased a significant amount of Western technology through licit, arms-length transactions, including tanks from the United Kingdom, aircraft from the Netherlands, and aircraft parts from the United States.[28] The prewar Soviet military DIB borrowed liberally from foreign technologies, often improving on foreign models, or at least improving production efficiency. Examples of appropriated technology include the T-18, a reproduction of French Renault FT-17 light tanks, produced from examples captured in the Russian Civil War.[29] In some cases the USSR negotiated licensing agreements for the production of parts or machine tools.[30]

Soviet success in appropriating and then transforming Western technology served as one of the key justifications for the postwar US creation of an elaborate system of export controls designed to prevent the USSR from acquiring Western innovations. The functioning of this system is discussed in greater detail in chapter 5, but in essence it sought to prevent the transfer of advanced technology to the Soviet Union. The Soviets countered this effort by developing a structured program for industrial espionage, targeting military and civilian technologies, a program that was fully operational by the late 1970s.[31] It was so structured, in fact, that it was demand-driven: twelve industrial branch ministries made requests for information to the Soviet intelligence community. Nine of the twelve branches focused on military technology, while the other three concentrated on civilian production. According to documentation acquired by the French government, the system acquired roughly five thousand items of hardware and eighty thousand technical documents per year.[32] The industrial branch ministries assessed and reported the usefulness of these acquisitions to the Soviet government. However, not all of the documentation involved industrial theft; like the United States, the USSR sought information about the strengths and weaknesses of foreign competitors.[33] Moreover, some of the documents and equipment acquired were available through open sources.

Defense analysts debated the impact of this industrial espionage during the Cold War. In the United States, the Department of Defense regularly claimed that Soviet military research depended on stolen Western technology, and called for tighter export controls. In the 1980s, for example, the Soviets used Western technology to assist in the production of armor-piercing shells, phased-array radars, infrared reconnaissance sets, computer systems, and submarine development technology.[34] The system did not work across the board, however; the Soviets enjoyed considerably greater success integrating foreign technology into their military production system than into the civilian economy. Given the rapid increase in the importance of dual-use technology in the 1970s and 1980s, this may have hamstrung not only the overall civilian economy but also the military sector itself, and in fact it may have contributed to the USSR's collapse.

THE SOVIET EXPORT SYSTEM

Soviet industry exported a great number of weapons to the rest of the world, with major customers including India, the People's Republic of China, Iraq, Egypt, Cuba, and most Eastern European countries, among others. By the late Cold War, the annual value of Soviet arms exports regularly reached $15 billion, some 40 percent of overall global arms transfers.[35] Both politics and the desire to earn hard currency drove Soviet export ambitions.[36] The Soviet Union exported arms to proxies and clients to ensure that the proxies had the military strength to further Soviet foreign policy goals, and that the clients remained in Moscow's political orbit.[37] Indeed, states that purchased Soviet arms and then strayed from the Soviet Union often faced serious problems maintaining their military capabilities. Indonesia, China, and Egypt all paid a price in terms of military effectiveness for defecting from the Soviet sphere.

The USSR also furthered both its economic and foreign policy goals by allowing Soviet firms to conduct licensing agreements with foreign firms, often supplying manufacturing kits and advisors to ensure successful implementation and spur local industry. However, most of these agreements were limited to the Eastern bloc, where the recipient regimes remained firmly under the Soviet thumb.[38] India and China were the major exceptions to this rule. Early in the Cold War, the Soviet Union licensed numerous systems to the People's Republic of China while also providing technical assistance to jump-start the Chinese defense industry.[39] This resulted in fighter aircraft, submarines, small warships, and a variety of small arms. The Sino-Soviet rift of the early 1960s effectively ended this relationship.

The Soviet Union also developed a series of licensing agreements with India, mostly in the aftermath of the Sino-Indian War of 1962, as the USSR attempted to use India as a counterweight to China. Immediately following independence, India acquired most of its arms from the United Kingdom and the United States, as befit its heritage as a former British colony.[40] After the 1962 war, however, India diversified its military technology supply portfolio and began to acquire not only Soviet weapons but Soviet industrial know-how. These agreements accelerated after the 1971 Indo-Pakistani War, which drove a wedge between Washington and New Delhi. The Moscow-Delhi relationship has persisted even beyond the end of the Cold War; Russia

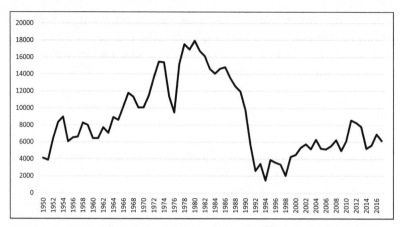

FIGURE 4.1. USSR/Russia arms exports, trend indicator value (TIV) by year, in billions of US dollars

Source: SIPRI, "Arms Transfer Database," https://www.sipri.org/databases/armstransfers

transferred a refurbished aircraft carrier to India, and Russia and India have worked together on several variants of the Brahmos cruise missile.[41]

The Soviets paid a price for the licenses they exported. The Eastern European countries often sought military export relationships with less developed countries (LDCs), cutting into the Soviet share.[42] In other cases they filled in for the USSR when tensions developed between Moscow and the LDCs. Erstwhile Soviet allies India, North Korea, and China undertook similar efforts, the latter two countries with far less respect for Soviet wishes and interests. This may have led the Soviets to curtail technology transfer to Eastern Europe and the rest of the world by the 1970s.[43] Figure 4.1 illustrates the arms exports of the Soviet Union and Russia using the Stockholm International Peace Research Institute's (SIPRI) Arms Transfer Database, which uses trend-indicator value (TIV) to operationalize the value of each arms deal. TIV measures the volume of the military resources themselves, rather than the transaction's financial value, calculated on the basis of the estimated production costs of the weapons traded to and from a country within a given year.

REFORM AND TRANSITION

Since the end of the Cold War, Russia has struggled to maintain the quality and size of its defense workforce. The dramatic downsizing of the defense industry in the 1990s meant that the surviving producers had little incentive to train new workers. Moreover, the increasing importance of the oil and gas industries attracted many of the engineering-oriented workers away from defense work. Combined with the relatively short life span of Russian men, this means that the Russian defense workforce has dwindled in size and quality. Indeed, several regular customers have complained of poor maintenance and workmanship on Russian weapons.[44] Today the Russian defense industry continues to consist of several large state-owned enterprises, with import and export activity coordinated through Rosoboronexport, a state agency regulating the transfer of military equipment and dual-use technologies.[45]

The end of the Cold War radically changed the environment of the Russian arms industry. Domestic demand for weapons collapsed; the Red Army and its sister organizations could not maintain the equipment they currently owned, much less purchase replacements or upgrades. Providing weapons and associated military technology to the Russian armed forces could no longer provide sufficient incentive for innovation, and the development of new systems stalled, or ended altogether. Consequently, the economic importance of exports soared.[46]

However, the outcome of the 1991 Gulf War produced a concurrent crisis that made it difficult for the Russian arms industry to increase exports. The course of proxy conflict during the Cold War had always left some question as to the relative quality of Western and Soviet equipment. Although Western proxies usually performed better in high-technology combat than their Soviet counterparts, observers could generally write the difference off to variables such as training, doctrine, and force employment. Moreover, Soviet weapons performed quite well in some key conflicts, including the Vietnam War (where Soviet MiG-21s outperformed larger, less maneuverable American fighters) and the Indo-Pakistani Wars, where Indian soldiers and pilots fought competently and effectively with Soviet equipment.[47]

In the Gulf War, however, Western weapons built and operated by the United States, the United Kingdom, and others cut through Iraqi forces with uncanny effectiveness. In both the air and ground wars, Western systems easily prevailed over their Soviet-built counterparts. Air attacks decimated Iraqi formations, the Iraqi air force refused to fight, and US Army spearheads utterly destroyed Iraqi defensive positions. To be sure, coalition forces had big advantages in training and doctrine. But justifiably or no, a perception developed that Western systems enjoyed a decided advantage over Soviet systems.[48]

The Russian DIB survived the 1990s primarily by massively increasing its exports to China. The Sino-Soviet split had severed the arms relationship between the two Communist giants in the 1960s, but a warming of relations in the 1980s opened the door to potential bilateral cooperation.[49] This warming trend accelerated with the end of the Cold War as Moscow abandoned its political qualms about exporting equipment to China and sent huge shipments of arms to the Chinese military, peaking at $2.5 billion annually during the 2000s.[50] It also produced kits for Chinese production, and engaged in some technology transfer.[51]

Today's Russian arms industry has not globalized to the same extent as the US arms industry. However, many Russian defense firms have important relationships with foreign supplies and partners, both in production and innovation. In the case of India, this has resulted from a diplomatic and military relationships that highlight joint innovation and production, though the nature of that relationship has changed since the end of the Cold War, given India's access to sophisticated American and European technology.[52] However, the multinational nature of the Russian defense industry also stems from patterns of innovation and production left over from the Soviet period. During the Cold War, the system of military production and innovation in the Soviet bloc encompassed Russia, the rest of the Soviet republics, and the Eastern European satellites. Some of these relationships ended between 1989 and 1991, but others have endured, especially among the former republics. Until recently, for example, Russia and Ukraine enjoyed a productive military relationship, with Ukraine manufacturing many key aviation components for delivery to Russian industry.[53] This relationship, along with a nascent relationship Russia had attempted to develop with France, fell victim to the 2014 Ukrainian revolution and the consequent Russian seizure of Crimea.[54]

POST-TRANSITION EXPORT

The Russian defense industry still produces a variety of systems developed in the Cold War. Russia continues to export all manner of weapons, but most involve incremental innovation on frames developed and deployed during the Cold War. The most lucrative exports include Kilo-class diesel-electric submarines, MiG-29 fighters, variants of the Su-27 "Flanker," T-90 tanks, and a variety of other vehicles and small arms.

New weapons include the Lada class submarine (an evolution of the Kilo design), the Borei class ballistic missile submarine (initially designed during the 1980s, though not laid down until 1996), and the T-90 tank, among others. Russia has seen less success with next-generation products. The Sukhoi PAK FA fifth-generation fighter promises, on paper, to compete effectively against Lockheed Martin's F-22 Raptor and F-35 Lightning II. Although Sukhoi expected the fighter to become the mainstay of the Russian air forces, it also sought joint production and export partnerships with India. Large-scale production of the PAK FA, however, may exceed the current capacity of the Russian arms industry and the Russian economy. Stealth aircraft require parts crafted within an exacting range of tolerances, and fielded prototypes have suffered from numerous problems. Significant tensions have emerged with India over partnered production of the aircraft, with Indian air force officers expressing open concern over the capabilities of Russian industry.[55]

Russia's export problems are not confined to the air. The Armata once represented Russia's most modern entry into the ground combat vehicle market, the vanguard of a family of armored vehicles that share a common chassis. Upon its appearance, some analysts suggested that the Armata could have a big impact on the international export market.[56] Uralvagonzavod, the developer of the combat vehicle, seemed to agree, and by 2015 it had begun hyping the prospect of sales to Egypt and various Central Asian states.[57] However, the Armata ran into significant cost problems.[58] Russia substantially cut back its expected purchase, and other countries balked at a high price tag for an admittedly excellent tank. Moreover, the export prospects of the Armata may face another problem: China has its own tank for export, the Norinco Type 90, which sells for less than the projected price of the Armata. The Armata design team has expressed concern about the possibility of Armata exports to China, on the grounds that China might reverse-engineer components and manufacture them domestically.[59]

INTELLECTUAL PROPERTY REFORMS

As discussed above, the Soviet Union bestowed an uncertain IP legacy upon Russia. Patent rights conveyed very little in terms of individual ownership or monopoly, and thus did not well serve the transition to a capitalist economy. Consequently, along with other sectors of the economy, the Russian intellectual property system underwent a massive overhaul in the Russian parliament in 1992 and 1993.[60] This legislation affected patent, trademark, and authorial rights, among others, and created a bureaucracy for managing and protecting these rights.[61] This first wave of legislation left much to be desired. Enacted in haste, the legislation often lacked clarity and coherence.[62] Reform faced significant obstacles, including a basic lack of understanding on the part of many actors within the economic system as to the functioning of essential IP concepts.[63] As Russia grew more familiar with the practice of IP protection, the need for further reform became clear.[64] Indeed, as Russian civilian and military firms became integrated into the broader global economy, and as the economy shifted toward private over state ownership (though less so in the defense field), IP reform became imperative. Realization of this problem led to an overhaul of IP law in 2007, which resolved many of the problems associated with existing law and introduced some new concepts.[65]

Russia's defense sector suffered from the lack of understanding of IP. As the relationship with China would eventually show, Russian firms displayed little sophistication with respect to management of their intellectual property.[66] They undertook few safeguards to prevent Chinese copying, either on the engineering or contractual levels.[67] Western firms may also have taken advantage of Russian desperation and lack of sophistication. In the early 1990s, Lockheed entered into a collaborative project with Yakolev, a state-owned aviation firm. Yakolev produced the Yak-38, one of only two VSTOL (vertical or short take-off and landing) fighters in the world. Yakolev had developed the Yak-141, a supersonic successor to the Yak-38, in the 1980s, but the Soviet Union collapsed before it could produce the aircraft in any numbers.[68] Unlike the Yak-38 or the Hawker Siddeley Harrier (the other VSTOL fighters), the Yak-141 used a rotating tilt mechanism that allowed it to use its main engine for both lift and thrust. A central turbofan provided power, which could be vectored depending on the aircraft's position.[69]

Lockheed and Yakolev failed to turn the Yak-141 into a commercially viable project, though they did build several working prototypes. A decade later, however, a turbofan system very similar to that of the Yak-141 appeared in the F-35B, the VSTOL variant of Lockheed Martin's Joint Strike Fighter.[70] Today, the F-35B is the world's only operational VSTOL fighter capable of supersonic speed. While the precise details of the collaboration between Yakolev and Lockheed remain hazy, there is little doubt that Lockheed engineers gained important experience from the joint project.[71]

Reform efforts continue today, as members of the Russian parliament seek to both improve the legal environment for domestic innovation and improve prospects for Russian companies operating abroad.[72] With respect to the latter, Russia has developed mechanisms for hiring Western law firms to protect Russian products abroad. Sergei Zuikov, director of a Russian patent firm, has commented, "Part of the state's plans [sic] is the creation of a system of accreditation of Western lawyers, who will defend Russian intellectual property abroad. Not all Western lawyers—even those who take . . . multimillion-dollar honorariums are professional enough and are interested in protecting the interests of their customers from Russia. In this regard, there is a need to hire those specialists, who will protect Russian intellectual property for a reasonable price on a regular basis."[73] Over time, however, Russia will need to increase domestic expertise on intellectual property protection.

China's National Innovation System

Political dysfunction and technological backwardness plagued China's defense sector during the twentieth century. Late in the Imperial Period, government and private actors began to borrow industrial technology from the West, and during World War I China became a significant industrial producer despite the chaos that continued to grip the country. However, China did not at this time achieve any lasting innovative success; its industry remained firmly behind the European competitors on the industrial frontier. The destruction associated with the Japanese invasion of 1937, along with the Chinese Civil War, left the Chinese Communist Party (CCP) nearly a blank slate when it took control of the country in 1949.

The Classic Chinese National Innovation System

In 1949 the Chinese defense industry produced little in the way of sophisticated military technology. World War II and the Chinese Civil War had destroyed much of the urban industrial base, and the Soviets had confiscated most of the industrial equipment the Japanese had brought to Manchuria in the 1930s. The dire economic situation that faced the People's Republic of China (PRC) in the wake of the revolution left little for investment in technological development.[74] To the extent possible, China established a defense industrial base on the model of the Soviet Union, built primarily around large state-owned firms.

As initially established, the Chinese DIB distinguished between the strategic weapons complex (nuclear weapons and their delivery systems) and the conventional weapons complex.[75] The former would have the latitude to engage in basic research, as well as a degree of protection from the vagaries of CCP politics. The latter would concentrate on production, imitation of foreign technology, and incremental improvement. The strategic complex managed to develop nuclear weapons with minimal foreign assistance in conditions of tremendous poverty. The conventional weapons complex produced a huge number of obsolescent planes, tanks, and ships, often a generation behind the industry standard. Both sides relied on state investment in large-scale state-owned enterprises.[76]

The advent of the Cultural Revolution in 1966 did not help matters. A combination of bureaucratic infighting, intra-CCP politics, and the broader ideological struggle for control of the communist world with the Soviet Union, the Cultural Revolution inaugurated a wave of anti-elite, anti-intellectual politics that afflicted both the civilian and the defense economies. Scientific and engineering expertise became suspect, limiting the potential for innovation and curtailing the development of the Chinese NIS. Although the most intense period of the Cultural Revolution lasted only two to three years, it had an enduring impact on the competitiveness and sophistication of the Chinese defense sector.[77]

INTELLECTUAL PROPERTY IN THE CLASSIC
CHINESE NATIONAL INNOVATION SYSTEM

Intellectual property considerations were virtually absent from the early Chinese national innovation system. A patent system did not, in any meaningful sense of the word, exist in either China's defense sector or its industrial sector more broadly. Unlike imperial Russia, which maintained intellectual property protection standards roughly in line with European contemporaries, the Qing Dynasty had no system of intellectual property protection, and no means of enforcing any system that it could import from the West.[78] Indeed, copying Western technology (and thus abrogating Western intellectual property) represented a central industrial development strategy for the Qing Dynasty, for Nationalist China, and for the PRC.[79] Rudimentary systems for IP protection were installed in 1903, 1912, and 1923, but disinterest and lack of enforcement capacity limited their effectiveness.[80] The PRC thus had little interest in borrowing an IP protection system (although it did borrow some specific procedures from the USSR), and no relevant model to borrow from.[81] Consequently, in the early history of the PRC, neither the nuclear nor the conventional sides of the Chinese NIS used patent protection.

While limited information from this period precludes a complete assessment, anecdotal accounts suggest that the lack of patent protection had severe negative effects on information sharing within the Chinese NIS, especially in the conventional arms industry.[82] Without a system for attributing credit for innovations, individual firms, labs, and research groups hoarded information even though the state owned all of them. This information hoarding limited the extent to which innovations could find their way around the NIS, not to mention the broader civilian economy.[83]

INDUSTRIAL ESPIONAGE AND THE CHINESE
NATIONAL INNOVATION SYSTEM

Most scholarly treatments agree that the Chinese defense industrial base has historically depended a great deal on illicit acquisition of foreign technology.[84] In the 1960s and 1970s, this primarily involved the appropriation

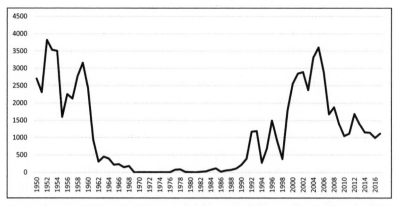

FIGURE 4.2. People's Republic of China arms imports, trend indicator value (TIV) by year, in billions of US dollars

Source: SIPRI, "Arms Transfer Database," https://www.sipri.org/databases/armstransfers

and repurposing of Soviet technology. From the 1980s on, the Chinese national innovation system has sought and acquired technology from Russia, the United States, and a variety of European targets.[85] For example, just on the aerospace side, the influence of foreign technology is clear. The J-10 fighter was based on the Israeli IAI Lavi and the US General Dynamics F-16; the J-11 fighter is a clone of the Russian Su-27; the JF-17 is a modern development of the Soviet MiG-21; and finally, the J-31 is widely reputed to rely heavily on technology associated with the US Lockheed Martin F-35.[86]

Chinese military technology firms acquire foreign technology through various means, both above and below board.[87] On the private side, Chinese firms operating abroad, and in partnership with foreign firms domestically, have access to an array of foreign technologies and production methods. Chinese students study in Europe, Australia, and the United States, becoming familiar with techniques developed in the world's most advanced research universities. They then bring specific expertise, as well as a variety of research and engineering techniques, back to the Chinese NIS.[88] China also acquires weapons and technology transfers through legitimate purchase; the acquisition of Russian fighter and jet engine technology began with the licit purchase of Russian Sukhoi aircraft.[89] But most notably, the People's Liberation Army has created a vast infrastructure for the appropriation of foreign intellectual property, especially in the defense sector.[90] Figure 4.2 illustrates the arms export behavior of the PRC.

As chapter 6 will discuss in detail, there are many practical obstacles to the successful theft of technology.[91] Individual bits of data, even sophisticated data associated with patents and trade secrets, mean little out of context. Would-be thieves need to know a lot about their targets, as well as a great deal about the subject matter involved. In order to produce useful innovation, the cyber-soldiers of the People's Liberation Army need to know where to direct their efforts, and what they need to look for. This requires close collaboration between the DIB, which knows what it needs, and the cyber teams, who know where to look. The responsiveness of these organizations remains among the most closely guarded secrets of Chinese intelligence, and there are few effective ways of measuring their impact.

The successful Chinese projects have required either significant production data (as was the case with the J-10), or actual physical examples to work from (as in the case of the Su-27). Despite this, Chinese efforts at reverse engineering have periodically hit production roadblocks; for example, a lack of access to Russian trade secrets associated with the construction of jet engines has led to considerable problems in China's own jet engine assembly.[92]

CHINESE EXPORTS

China has long exported military equipment to the world, but for most of the postwar period this has involved second-rate, low-technology weapons. These sales sometimes went to states that could not afford similar Soviet systems, or which had found themselves in some sort of political difficulty with Moscow. Albania, for example, purchased extensively from China after it split from the Warsaw Pact.[93] In part because of the rudimentary nature of the weapons exported, and in part because of the nature of the weapons delivered, intellectual property concerns rarely arose in these relationships.

One interesting exception to this rule came in the early 1970s, when the PRC sold a squadron of J-7 interceptors to the United States. The Sino-Soviet split had taken place prior to the full development and licensing of the MiG-21 to the PRC, but in an effort to heal the political divide the Soviets had exported plans and a working model to China in the early 1960s.[94] The Soviets ceased cooperation after the signing of a licensing deal, forcing the Chinese to proceed with limited technology and expertise.[95] Nevertheless, the Chinese persevered, eventually producing the J-7 for domestic use and the F-7 for export. In the early 1970s, China delivered a group of these aircraft to the United States for use in aggressor training, presumably violating the licensing agreement with the Soviet Union.[96] It eventually exported the F-7 to fifteen different countries, many of which still fly the fighter.[97]

REFORM AND TRANSITION

Patent protection proved a controversial element of Deng Xiaoping's economic reforms, but China passed its first comprehensive patent legislation in 1985.[98] These reforms, based around the German model of patent protection, tended to favor foreign investors rather than domestic producers. In the late 1990s and 2000s the CCP pushed a major set of reforms through the defense industry.[99] The largest, most important firms remained state-owned, but were forced to reform in order to increase efficiency and responsiveness, and to reduce cost.[100] The government tried to create a competitive environment by splitting firms and setting them against one another, and by stepping up purchases from Russia.[101] Indeed, unlike the DIBs of the United States and Europe, the Chinese defense sector has enjoyed consistent increases in domestic procurement funding since the early 1990s, allowing it to invest in innovative technologies and production facilities.

This is not to say that the Chinese DIB can compete in technological terms with the most innovative firms in Europe, Japan, and the United States. Most firms in the Chinese DIB have concentrated on incremental innovations, adapting newly developed and acquired technologies to old platforms in small-batch construction. Chinese firms have also specialized in what scholars describe as "architectural" innovation; innovations that shift and repurpose existing technologies in new forms, hopefully with emergent qualities.[102] Architectural innovations can reap tremendous rewards in military technology. The world-beating battleship HMS *Dreadnought*, for example, represented an architectural innovation, combining unrelated innovations in gunnery, armor, and propulsion. Similarly, the Df-21 carrier-killer antiship ballistic missile, a weapon that some have argued can unsettle the balance of power in the Pacific, repurposes existing communications, geolocation, and missile technology into a new form.

China's DIB has not globalized to the same extent as that of the United States or even Russia. China never had a collection of satellite states that formed an integrated system of military production, and consequently Chinese innovation and weapons manufacturing has remained firmly within its borders. The exception to this rule came in the 1970s and 1980s, when China's defense industries enjoyed a degree of collaboration with American and European partners. Motivated both by profit and by the desire to weaken the USSR, Western firms worked to improve Chinese military capabilities, especially in electronics. However, this period ended with the imposition of sanctions following the 1989 Tiananmen Square massacre.[103]

The reforms of the 1980s and 1990s pushed the defense industry into the civilian economy, often unwillingly. Firms often had to restructure in order to produce goods for the civilian market, and this sometimes reduced efficiency and innovative capacity. However, the restructuring also tended to improve the internal operation of firms, familiarizing them with the prospects of the civilian market.[104] More recently, the Chinese defense industry has moved toward the Western model, with strong ties developing between large state-owned defense enterprises and smaller technology firms. In the Chinese case, the lack of access of traditional defense providers to the wider world of military technology makes it even more important for Chinese defense firms to work with their civilian counterparts.[105] Until 2002, private Chinese firms could not compete with state-owned enterprises for contracts with the People's Liberation Army. Between 2002 and 2005, regulatory reform opened up the PLA's system of contracting, theoretically allowing private firms to compete.[106] However, the process of creating a regulatory system for managing such bids has proceeded slowly. State-owned enterprises, which have powerful connections within both the People's Liberation Army and the Chinese Communist Party, retain huge bureaucratic advantages over their private competitors, even when the private firms can aggressively bid.[107]

As with the United States and most Western countries, the Chinese government realized in the 1990s that civilian and military technological needs overlapped, especially in high-end electronics, and that an increased emphasis on research would benefit both the civilian and the military sectors.[108] The problem lies in the need for a regulatory framework that will allow the integration of civilian and military research—one that ameliorates distrust between civilian and military researchers while also providing incentives for collaboration.[109]

China has taken some steps toward reversing this isolation by globalizing its defense industry, mainly by allowing foreign investment in sensitive defense-related firms.[110] This grants the firms access to capital, bureaucratic expertise, and some technology. The biggest global influence on the Chinese defense sector, however, has come not through partnerships but through espionage. Chapter 6 covers this phenomenon in greater detail.

POST-TRANSITION EXPORT

The increasing sophistication of the Chinese DIB could make it more competitive for higher-tech equipment, but Chinese firms have had trouble breaking into some of the more lucrative markets. For example, the Chengdu Industry Group may be on the verge of some success with the JF-17 fighter, though as this aircraft strongly resembles an updated MiG-21, it does not serve to demonstrate cutting-edge technological innovation. Thus far, the only customers are Myanmar and Pakistan, but rumors suggest that Nigeria, Egypt, and Argentina may also have some interest.[111] Nevertheless, during the period of reform the overall value of China's military exports has generally increased, as can be seen in figure 4.3.

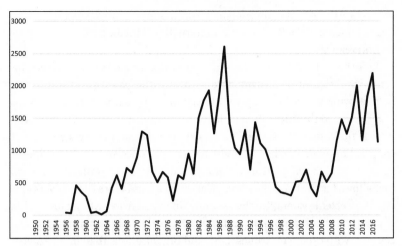

FIGURE 4.3. People's Republic of China arms exports, trend indicator value (TIV) by year, in billions of US dollars

Source: SIPRI "Arms Transfers Database," https://www.sipri.org/databases/armstransfers

INTELLECTUAL PROPERTY REFORM

A combination of the move to an export-driven economy, an increase in the importance of the civilian-technology sector, and an interest in increasing the level of innovation across the economy drove intellectual property protection reform in the 1990s and 2000s. Part of the impetus for reform came from US pressure. From the 1990s on, the United States has consistently and vigorously critiqued the Chinese government for its relaxed attitude towards the protection of intellectual property. Copyright has come under the most scrutiny; US films, television, and music commonly appear in China with little attribution and no profit for artists or producers.[112] In addition, the US government has critiqued China for ignoring patent protections and for appropriating Western trade secrets.[113] Indeed, intellectual property disputes have become the central tension in the US-China trade relationship. President Donald Trump has made Chinese IP theft a central plank in his critique of China.[114]

This pressure, along with the influence of international intellectual property regimes, has undoubtedly had an effect on Chinese behavior.[115] In 1998, the PRC established its State Intellectual Property Office, updating older institutions that could not manage foreign IP claims.[116] This strategy represented an enormous step toward bringing China into formal accord with developing international IP norms. The Chinese government also established a bureaucracy for patent protection, though Beijing's interest in creating the bureaucracy stemmed both from international pressure and from a desire to increase the rate of innovation in the domestic economy.[117] Along the way, China joined most of the important international intellectual property protection regimes.

Reforms to the system of IP practice in the Chinese DIB have helped incentivize information sharing, and an across-the-board regulatory effort helped bring many firms up to international standards.[118] These reforms also facilitated the integration of dual-use technology into the Chinese NIS. As was the case in the United States, Chinese civilian firms worried about cooperation with the defense industry, out of concerns over state appropriation of technology. China employs a dual-track IP protection system that allows for the use of technological innovations before those inventions enter the patent system. This leaves individuals and firms without protection and at risk of appropriation, and thus less interested in sharing their discoveries with other companies, labs, and organizations than in a Western-style patent system.[119]

Moreover, a culture of patent application has yet to take full hold in the Chinese science and technology communities, as inventors often fail to take advantage of even the limited protections offered by the IP system.[120] In part this has resulted from the underdeveloped nature of the Chinese IP system. As noted above, China is in the process of developing a bureaucracy for the protection of intellectual property, but this development has thus far concentrated on broad directional goals rather than on regulatory detail.[121]

China also has yet to resolve the IP problems posed by state ownership. Because of the prominent role of state-owned enterprise in the defense sector in both industrial production and research, even innovations that reach the patent stage often remain the property of the state. This strips financial incentive from researchers working on military technological innovations, especially when those researchers work outside the state (for private firms, private research labs, or universities).[122]

Weak IP protections also limit technology transfer from the defense sector into the civilian economy. China's state-owned defense firms, and the constellation of research labs operated by the People's Liberation Army and other organs of the Chinese government, often develop dual-use technologies. However, the PLA's incentive for sharing whatever innovations do not compromise state secrecy depends upon the health of the IP system, and thus far the military has not demonstrated much trust in the robustness of legal protection.[123] As in the United States, the bureaucracy for protecting state secrets, and thus preventing the release of sensitive technology, has proven difficult to navigate.[124] The size and complexity of the state-owned component of the Chinese military-industrial complex makes it difficult for anyone to assess the breadth of technologies for potential civilian use.[125]

In short, the Chinese defense industrial base has matured along with the larger Chinese economy. This maturation, along with exposure to the international economic regime, has forced the Chinese government to adopt more complex and restrictive rules of information sharing. This adoption remains uneven, but it seems to guide the behavior of an ever greater portion of China's domestic firms.

The Second Tier: South Korea and the Hyundaization of the Arms Industry

At the end of the Korean War, the Republic of Korea (South Korea) possessed little in the way of advanced industry or industrial know-how. Through an export-led growth strategy with significant state investment, and with the economic and technological help of the United States, South Korean industry today has reached the pinnacle of several sectors, from shipbuilding to automobile construction to telecommunications. And although South Korea equipped its military primarily with American arms through most of the Cold War, today its defense industrial base has developed a prominent export profile.

The South Korean defense industry is sometimes mentioned as part of a phenomenon referred to as the "Hyundaization" of the global arms export business. The reference is based on the idea that the Hyundai automotive conglomerate has succeeded globally by producing "good enough" automobiles, while not competing on the high end.[126] The argument suggests that medium producers—South Korea, Brazil, Turkey—can undercut the dominance of the United States and Europe in the most sophisticated arms by providing equipment nearly as good for a lower price. While analysts of the global defense industry have warned of the rise of the second tier for quite some time, these midrange producers have yet to seriously undercut the biggest players.[127] Part of the reason may be that the United States has legal tools (offered by the export control regime and the intellectual property regime) for preventing South Korea and similar states from exporting arms.

South Korea is one of only a few countries involved in both the import and the export of high-technology military equipment. Much of its most advanced equipment comes from the United States, with some systems supplied by European providers. For example, the bulk of its air force consists of F-15s and F-16s. South Korea's large navy, mostly built by its bustling shipbuilding industry, uses US technology. The Aegis air defense system, for example, provides the primary defensive weaponry for its larger destroyers. One of South Korea's main battle tanks, the K1, is based on the US M1A1, albeit with significant modifications.[128] The second, the more modern K2, uses indigenous technologies in order to avoid licensing issues associated with export of US military equipment.[129]

Intellectual Property Protection in South Korea

As a member of the World Trade Organization since 1995, South Korea has adopted most of the major international legal requirements for IP protection.[130] It joined the World Intellectual Property Organization in 1996.[131] The US–South Korea trade agreement of 2012 further boosted mechanisms for the protection of IP.[132] During the 1990s, South Korea briefly became notorious for lax enforcement of IP norms, its membership in major international organizations notwithstanding. Some scholars have associated this lack of compliance with enduring cultural factors.[133] However, in the ensuing years South Korean compliance has shifted markedly for the better, suggesting that enforcement and government commitment play a major role in the functioning of IP law.[134]

South Korea entered the export-control regime relatively late, as it did not have a functional defense industry until the 1980s. By this time, however, a robust consumer electronics industry had developed which could potentially export dual-use systems of some concern to the United States. Indeed, during the 1980s South Korea specifically sought a larger trade relationship with the Soviet Union and the rest of the communist bloc.[135] In 1987 it formally acceded to the US export control regime.[136] Its export control system now functions in terms broadly similar to that of the United States, granting differences that result from the different positions of the two countries. In terms of dual-use technology, South Korean firms play a large role in global electronics. Samsung, in particular, occupies a central place in the Korean economy, exporting electronics, computers, and other technology around the world.[137] Between 2011 and 2018, Samsung became engaged in a wide-ranging series of IP disputes with Apple about the design of tablets and smartphones.[138] The case, litigated across several different countries, was seen by some as important to the development of international IP jurisprudence.[139]

The US–South Korea Free Trade Agreement (KORUS) of 2012 included provisions designed to help solidify South Korea's commitment to intellectual property protection.[140] Even after KORUS, however, US business representatives continued to criticize South Korea's IP protection practice. The US pharmaceutical industry, always particularly sensitive to IP violations, complained in 2017 that the South Korean government was unfairly regulating against US patents, assigning highly valuable pharmaceuticals to categories that received little protection. These complaints provided grist for the Trump administration's aggressive approach to trade negotiations with South Korea.[141]

Innovation and Espionage

The South Korean defense industry consists of firms jointly held by private groups and by the government, and has a complex set of connections to both the civilian economy and to foreign defense producers. Big players include Hanwha (a defense electronics firm spun off from Samsung), Korean Aerospace Industries (KAI), the shipbuilders HHI, DSME, and HHIC, and the munitions and electronics firm LIG Nex1.[142] The state has a significant ownership stake in several of these companies; 26 percent in KAI, for example.[143] South Korea also carries on a robust arms export business, as would reflect its leadership position in many high-technology categories. In fact, it was the world's thirteenth largest arms exporter in 2013, dealing mostly in warships and aircraft but also in missiles and artillery.[144] Because of South Korea's position within the international arms trade, its export control mechanism does much of its work negotiating with other (primarily American) systems of licensing.[145]

South Korea's defense industrial base commonly undertakes joint projects with US firms, and less often with European firms. Such joint efforts can provide a country with critical technologies, but such ventures come at a significant cost. For example, he KF-X project demonstrates many of the pitfalls of the internationalization of the arms industry. In the late 2000s, South Korea initiated the KF-X project as a means of developing an indigenous fifth-generation stealth fighter, potentially competitive with the US F-35, the Soviet PAK FA, or the Chinese J-31.[146] KAI pursued this project in cooperation with Lockheed Martin, which had developed considerable expertise with stealth technology as a result of the F-22 and F-35 projects. The deal involved the purchase of forty F-35 joint strike fighters, along with twenty-five technology transfer projects. This fulfilled South Korea's need for a fifth-generation fighter to replace its aging F-15s and F-16s, and promised to revitalize the South Korean aviation industry.

In 2015, South Korea requested the transfer of the twenty-five projects in order to facilitate the further development of the program. Despite Lockheed Martin's willingness to sell the technologies and send engineers to Korea to facilitate the technology transfer, the US Defense Acquisition Program Administration refused to grant an export license for four of the most important technologies, presumably on the grounds that the technologies were too sensitive for export even to a close ally.[147] They included an active electronically scanned radar, the infrared search-and-rescue systems, the electro-optical targeting pod, and the radio frequency jammer.[148]

The lack of these technologies will almost certainly have a long-term detrimental effect on the KF-X program, which was intended to provide 120 fighters for South Korea, and another 80 for Indonesia, between 2025 and 2030.[149] South Korea can acquire some of the technologies from other sources, such as European defense firms, and can develop some on their own, but only with considerable delay to the overall project.[150] And even if it manages to acquire or develop the four technologies, it cannot export the KF-X without US cooperation. As will be discussed in chapter 5, the US system of export control gives the US government a veto over how foreign countries can transfer US technologies. The presence of significant US technology in the KF-X project makes it difficult for South Korea to sell the fighter without a US go-ahead. In 2015, for example, the United States vetoed the transfer of the KAI T-50 jet trainer / light attack aircraft to Uzbekistan, on the grounds that such a sale would violate US technology rights.[151] KAI developed the T-50 in collaboration with Lockheed Martin, and though the fighter does not include any extremely sensitive technologies, the fact of collaboration gives the United States enormous influence over the marketing and export of the aircraft.

The reticence to transfer the technologies to South Korea may result from long-term concerns on the part of US officials over the theft of other military technology. As aforementioned, part of South Korea's strategy for developing a robust arms export involves a strategy of indigenous technological development, specifically to avoid complications with foreign licensing and thus avoid a US veto on such transfers as the T-50 to Uzbekistan.[152] The K2 main battle tank has already won one export customer (Turkey, in a deal that included technology transfer), and South Korea hopes it will win more competitions in the future.[153] The KW1 Scorpion armored personnel carrier uses a strategy similar to that of the K2, consisting almost entirely of indigenous technologies in order to avoid licensing problems.[154]

US officials have reportedly expressed concern, repeatedly, that this industrial strategy involves the quiet replication of US defense technologies to avoid licensing concerns. In particular, some US officials suspect that the South Koreans have appropriated fire control technology for the K2 tank, as well as structural and propulsion technology for the Haesung anti-ship missile.[155] The United States (which has reportedly expressed these concerns directly to South Korea in secret talks) worries both about the potential that South Korea might transfer these technologies to a hostile third party, and also that South Korean exports could undercut US exports in the same sec-

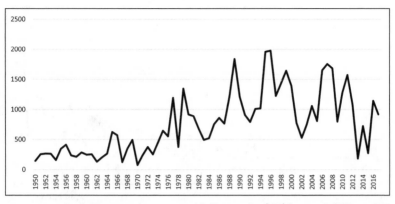

FIGURE 4.4. South Korea arms imports, trend indicator value (TIV) by year, in billions of US dollars

Source: SIPRI, "Arms Transfer Database," https://www.sipri.org/databases/armstransfers

tor.[156] Unlike many American export partners, South Korea has the indigenous industrial and technological base to appropriate and replicate the most advanced US technologies.

Concerns over theft have not dissuaded US or European producers from additional alliances with South Korean firms. In addition to South Korean companies' partnerships with Lockheed Martin on both the KF-X and the T-50, the South Korean shipbuilding firm Daewoo worked collaboratively with the German firm Howaldtswerke-Deutsche Werft (HDW) to produce under license the Type 209 submarine, which South Korea has agreed to export to Indonesia.[157] So, while the quality of the South Korean products may not be high enough to attract US and European buyers, that does not stop both South Korean and Western firms from pursuing manufacturing partnerships. Figures 4.4 and 4.5 describe the arms import and arms export behavior of South Korea.

Conclusion

The structure of a national innovation system, and consequently of a defense-industrial base, depends on many things other than intellectual property protection. Government ownership, the nature of relations between government and private defense firms, and the position of a state within the international arms export market all matter for how a DIB functions.

That these factors are critical does not mean that there is a single approach a country can adopt for the creation and operation of an effective national innovation system. Even for China and Russia, the former of which built its DIB upon the pattern set by the latter, IP protection practice varied

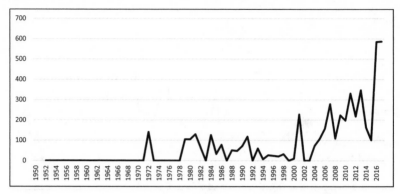

FIGURE 4.5. South Korea arms exports, trend indicator value (TIV) by year, in billions of US dollars

Source: SIPRI, "Arms Transfer Database," https://www.sipri.org/databases/armstransfers

considerably. Moreover, the IP protection systems of both varied over time, with major change coming in the 1990s and 2000s. Despite being radically different in design and execution, in the United States the relevance of IP law to defense procurement and the practices of the Department of Defense has varied over time.

The South Korean DIB, by contrast, has developed in full exposure to the international economy, and in particular the international intellectual property regime. Along with several other sectors of its industrial economy, its DIB has become internationally competitive, even as American export control restrictions continue to limit its international impact. These restrictions have helped shape South Korea's DIB, pushing it to selectively adopt autarky in service of export goals.

Some common themes emerge, however. In all three cases, the importance of intellectual property protection has grown over time. In the Soviet and Chinese cases, this happened in part because of dissatisfaction with the pace of innovation, especially relative to the West. In all three countries, an increase in IP protection has come in response to pressure from the emerging international intellectual property regime, and from the need to manage the evolving nature of public-private partnerships in the defense sphere.

The future success of these national innovation systems depends to a great extent on their capacity to take advantage of dual-use technologies, and to manage the relationship between public and private actors. Intellectual property protection lies at the center of this effort. Establishing a bureaucratic system that can best use IP law to ensure military access to privately held IP, ensure sufficient protection to allow the transactional sharing of IP between private firms, and protect all IP from outside actors is the key problem for the modern research, development, and acquisition system.

5: INTELLECTUAL PROPERTY
AND THE ARMS TRADE

Vignette: Selling the Flanker

In the late 1960s, the Soviet Union became aware of an American air superiority fighter project that would eventually become the F-15 Eagle. At the time, the USSR relied on the MiG-21, the MiG-23, and the MiG-25 for air superiority and interception missions, but the Soviets expected that the new American fighter project would exceed all three in capability.[1] Consequently, the Soviet general staff solicited designs from Sukhoi and Mikoyan for fighters that could match the F-15's expected performance. Mikoyan produced a lightweight fighter that became the MiG-29, while Sukhoi produced a heavier aircraft, eventually classified as the Su-27, NATO code "Flanker."[2]

The first Flankers entered service in 1985, but production problems kept their numbers low until the early 1990s. At that point, the collapse of the Soviet Union significantly reduced the expected production run.[3] Nevertheless, the Flanker eventually proved successful, with performance exceeding design expectations. Moreover, the basic Flanker frame has proven remarkably flexible for upgrade, and, as a result, the Flanker has become the platform of choice for discerning, budget-conscious fighter customers. Russia has exported the Flanker around the world, but India, China, Ukraine, and Russia itself remain the most important operators. Variants of the Flanker include the Su-30 multirole fighter, the Su-33 carrier-based fighter, the S-34 fighter-bomber, the Su-35 air superiority fighter, the Shenyang J-11 air superiority fighter, the Shenyang J-15 carrier-based fighter, and the Shenyang J-16 fighter-bomber. Both Russia and China continue to build Flanker variants, though problems have recently emerged with the Russian production chain, as Russia depends on Ukrainian suppliers for some components and the 2014 conflict between Russia and Ukraine over Crimea disrupted this process. Table 5.1 lists the variants and users of all variants of the Flanker family.

Little in this history distinguished the Flanker from earlier Soviet fighter aircraft. The USSR exported most of its tactical aircraft widely during the Cold War, sometimes using the airframes as loss leaders for engines and

TABLE 5.1. Sukhoi Su-27 variant transfers

Variant	Source	Operator	Number	Use
Su-27	Russia	Angola	8	Multirole fighter
Su-27	Russia	China	75	Multirole fighter
Su-27	Russia	Eritrea	9	Multirole fighter
Su-27	Russia	Ethiopia	12	Multirole fighter
Su-27	Russia	Indonesia	5	Multirole fighter
Su-27	Russia	Kazakhstan	30	Multirole fighter
Su-27	Russia	Russia	412	Multirole fighter
Su-27	Russia	Ukraine	70	Multirole fighter
Su-27	Russia	Uzbekistan	34	Multirole fighter
Su-27	Russia	Vietnam	12	Multirole fighter
Su-27	Russia	United States	2	Multirole fighter
Su-27	Russia	Belarus	28	Multirole fighter
Su-30	Russia	Algeria	44	Multirole fighter
Su-30	Russia	Angola	18	Multirole fighter
Su-30	Russia	China	100	Multirole fighter
Su-30	India	India	140	Multirole fighter
Su-30	Russia	Indonesia	11	Multirole fighter
Su-30	Russia	Kazakhstan	4	Multirole fighter
Su-30	Russia	Malaysia	18	Multirole fighter
Su-30	Russia	Russia	66	Multirole fighter
Su-30	Russia	Uganda	6	Multirole fighter
Su-30	Russia	Venezuela	24	Multirole fighter
Su-30	Russia	Vietnam	24	Multirole fighter
Su-30	Russia	India	90	Multirole fighter
Su-33	Russia	Russia	33	Carrier fighter
Su-34	Russia	Russia	84	Strike fighter
Su-35	Russia	Russia	40	Air superiority fighter
J-11	China	China	253	Multirole fighter
J-15	China	China	15	Carrier fighter
J-16	China	China	24	Strike fighter

other components. Expecting a short, sharp war against NATO, the Soviets did not design their planes to last, a reality sometimes lost upon their customers.

The real complications emerged following the export of Flankers to China. The Sino-Soviet split had largely ended the USSR-China arms trade in the 1960s, but the end of the Cold War reopened the relationship. In the early 1990s, China purchased forty-eight Su-27s from Russia in two groups of twenty-four. Russia followed up this sale by delivering one hundred Su-30 MMKs (a long-range strike variant) to the PLAAF and the PLAN.[4] In 1995, Russia and China came to agreement for the licensed production of two hundred Su-27 derivatives by Shenyang, a major Chinese aviation firm. Russia agreed to transfer production airframe kits to China, with final assembly within the PRC. The deal included IP rights to some of the technologies associated with the aircraft, including production trade secrets.[5]

The resultant aircraft became the Shenyang J-11, an aircraft that NATO refers to as the "Flanker B+." At some point in the early 2000s, however, the contract fell into dispute as Russia accused China of improperly fitting Chinese avionics and electronic equipment into the Flanker airframes.[6] Russia claimed that this would undercut long-term maintenance and supply contracts, and that it might allow China to export the J-11, in direct competition with Russia's own export ambitions and in violation of the contract.[7] Also, Russia disputed the legality of China's development of the J-15, a carrier capable variant of the J-11 that could compete directly with the Su-33, Russia's carrier-variant of the Flanker family.[8] Russia suspended delivery of kits at around one hundred, although China continued to construct the J-11 and its variants. The deal dissolved in acrimony, with a concurrent deal for the delivery of Il-76 transport aircraft also disintegrating.[9]

For its part, Shenyang argues that it has substantially improved the J-11 by including Chinese technology, and that the inclusion of these technologies did not invalidate the agreement with Russia. China clearly can now produce the aircraft on its own, with a variety of key modifications (the radar systems on the J-11D, the latest variant, exceed the capabilities of their Russian equivalents).[10] But Chinese copies have had their own problems. While China's aircraft industry has proven capable of replicating the Su-27 airframe, it has run into significant problems with engine production.[11] The WS-10 jet engine, for example, burns out after thirty hours, one-tenth the time of the same engines built in Russia. Chinese engines also deliver less thrust than their Russian counterparts, providing jets with less power. These problems have endured even as other aspects of China's aviation industry have moved forward.[12]

Regardless of which side bears the blame, the Flanker fracas created serious diplomatic problems between Russia and China. Until 2016, Russia and China could not come to agreement on the sale of additional Flankers, notwithstanding interest on both sides. China remained interested in the Su-35, Russia's most advanced version of the Flanker, but Russia was reluctant to sell, given its concern over China's past behavior. Contract negotiations focused on the delivery of complete aircraft, without production in China, and with a set of anticopycat and antireplication clauses. However, because the Su-35 carries an even more advanced version of Russia's best engine, Russia continued to balk unless the deal was substantial enough to guarantee a profit.[13] Eventually, Russia and China signed a deal for the transfer of twenty-four Su-35s beginning in December 2016.[14]

More broadly, these problems cooled Russian enthusiasm for a share of the Chinese arms market.[15] Russia has hesitated to export submarines, ballistic missiles, and air defense systems to China, largely out of concern that China might copy the equipment and compete with Russian products. More recently, engineers associated with the Russian Armata family of armored vehicles expressed concern about the prospect of exports to China.[16] Russia expected the Armata, an advanced chassis with configurations that run from main battle tank to infantry fighting vehicle, to become its primary mechanized vehicle in the 2020s. However, because of the cost of the vehicle and the relative success of the T-72 main battle tank and its variants, Russia decided to curtail production of both the domestic and the export versions of the Armata.[17]

China represents the best and worst kind of customer for Russian weapons. Beijing has the financial resources to purchase Russia's most sophisticated systems, but has sufficiently advanced technical and industrial capabilities to pose a significant IP threat. This is a classic IP dilemma, generated by the organization of Russian and Chinese industry. The iterative development of the Flanker represents a series of sustaining innovations built onto a single airframe, and we would expect large-scale, state-owned defense industries to excel at just this kind of innovation. The maturity of the technology, however, makes it difficult for Russia to keep the genie in the bottle. The Flankers have been on the market long enough that China can produce the airframes without Russian assistance or Russian license, and can equip them with Chinese electronics and other technology. Consequently, Russia depends on trade secrets and industrial practice to maintain an edge over China. For now, these trade secrets, as well as other industrial practices, allow Russia to maintain an advantage in engine construction; but that advantage may not last.

In short, Russia and China have an intellectual property law problem.

Russia would like to sell its most advanced military equipment to China, in no small part because it needs foreign sales to keep its military-industrial complex modern and competitive. China wants to buy Russia's most advanced military equipment, both in order to increase its military capabilities and because it wants access to Russian technology. China worries, though, that Russia will renege on a deal if China engages in what Beijing characterizes as "incremental" improvements to Russian technology. Russia worries that the Chinese will make cheap copies of the weapons they sell them and undercut their efforts on the global military technology market, especially in the developing countries that account for most of the Russian sales. However, the lack of enforceable intellectual property rights in either country, as well as an applicable international IP protection regime, prevents the two countries from reaching a deal.

This demonstrates the limits of an "arms-length" approach to the protection of IP. Russia can protect its military intellectual property simply by refusing to sell to China, but this tactic has its limits, especially if Chinese firms use industrial espionage or reverse engineering of third-party purchases to acquire the technology. A bilateral and multilateral framework for the enforcement of Russia's IP rights could free both countries from their concerns and increase the security of each. The prospects of both countries respecting (and relying on the other to respect) such an agreement may seem farfetched, but this is precisely the kind of architecture that the United States has attempted to construct for all forms of intellectual property.

Introduction

This chapter investigates the impact of intellectual property on the international arms market. The modern defense industrial base (DIB) has, since its inception, catered to both domestic and international customers. In the nineteenth century the major arms producers of Europe began exporting military equipment, including ships, small arms, and artillery, to smaller powers in Europe and around the world.[18] The importers lacked either the industrial technology to produce their own modern equipment, or represented a market too small to justify the development of a domestic arms industry of their own. In fact, large countries often fought wars almost entirely with equipment purchased from Europe; the Sino-Japanese War, for example, was fought almost entirely with warships acquired from European producers.[19] After World War II and decolonization, American, Soviet, and European arms found their way to newly independent governments around the world.[20]

Exporters have an incentive to control the intellectual property associated with their equipment in order to safeguard the effectiveness of their weapons, and to maintain the competitiveness of their exports. Importers, on the other hand, have different preferences: although some are content with the purchase of finished equipment and will contract with the seller for maintenance, replacements, and spares, other buyers want the technology and intellectual property associated with the weapons they purchase. Some may want to produce the weapons themselves.

At this point, intellectual property law intervenes. In general, importers and exporters have different interests, and the contractual agreements they arrive at take place within a framework structured by intellectual property law. Sometimes, weapons transfers convey no IP beyond that present in the weapons themselves; a one-off sale of small arms, for example. At other times, export arrangements will include provision for technology transfer associated with reducing the maintenance and repair costs associated with the weapons; machine tooling and instructions for construction spare parts, for example. Other agreements lay the foundation for joint production, which involves the transfer not only of various patented technologies, but also, necessarily, of trade secrets associated with manufacture.

National governments have generally supported arms sales, for both economic and political reasons. On the economic side, exports often generate a very high percentage of a DIB's total arms sales.[21] In the post–Cold War era, for example, some analysts estimated that Russian arms exports constituted around 85 percent of the total activity of the Russian defense industry.[22] Many of the major European producers, such as Sweden and France, have similar, although less pronounced, patterns. Indeed, only a few DIBs (in the United States, China, and perhaps India) expect to cater primarily to domestic customers. In addition to initial sales, the transfer of military equipment generally comes with a set of long-term agreements for maintenance, parts, and upgrades.[23] These long-term contracts are sometimes more lucrative than the initial sales.[24]

On the political side, arms exports can help ensure a long-term political relationship between patron and client state. The need for service and updates can lock a client into a relationship with a patron. Advanced systems often require a long-term advising presence as well, in order to ensure the proper employment and maintenance of the weapons. Arms transfers can also increase the strength of proxies, and help secure the terms of an alliance.[25]

But while arms deals bring clear benefits, they also have pitfalls. Selling a weapon to another state carries the straightforward threat that buyer will use the weapon against the exporter. During the Falklands War, Argentina operated several British-designed and -built Type 42 destroyers and a

Colossus-class aircraft carrier, all sold to Buenos Aires in happier times.[26] The threat of regime change can also make sellers leery of the transfer of some systems. For example, the Islamic Republic of Iran Air Force inherited several dozen F-14 Tomcats from the military of the Shah, and is currently the only air force in the world operating the aircraft. The reluctance of the United States to expedite the transfer of F-16 Vipers to the Iraqi Air Force has come in part from concern over the future of the Baghdad government. Indeed, the desire to control exports played a role in developing intellectual property protection at the dawn of the modern defense industrial base. As Katherine Epstein has pointed out, part of the US government's interest in acquiring property rights in torpedo technology came from an interest in preventing the exportation of that technology to potential competitors.[27] In fact, the US government was so concerned about keeping torpedo technology stateside that they used anti-espionage statutes to claim partial ownership of it.

States also face a variety of political obstacles to arms exports. Allies can have a significant effect on arms export decisions, with one government pressuring another not to export advanced systems. The United Kingdom, for example, places a great deal of pressure on other European countries to restrict exports to Argentina, and the United States does the same to maintain restrictions on European and Israeli technology exports to China.[28] The sale of advanced fighter aircraft to Gulf states invariably causes problems between Israel and the United States, although generally not serious enough to cancel the deals.[29]

In addition to political obstacles, states can also face legal obstacles to exporting weapons. After the Tiananmen Square massacre in China, for example, the EU imposed an arms embargo that has lasted until the current day.[30] The United Nations has imposed a variety of arms embargos in its history, usually against conflict areas to prevent escalation. On a unilateral basis, the United States has imposed numerous arms embargos (legal prohibitions against the sale of weapons) against unfriendly or suspect states, based on a wide variety of justifications. Such unilateral embargos can have greater impact than unilateral sanctions because of the market position of big suppliers (the United States, Russia, China, and France dominate much of the arms export market) and because of the long-term nature of a military relationship; the suspension of ties can leave existing equipment nonfunctional and useless.[31]

Finally, international legal agreements have tightly restricted the transfer of some kinds of military technology. The Missile Technology Control Regime (MTCR) includes many of the world's most advanced military producers and regulates the export of systems associated with ballistic missiles.

Because they cannot buy ballistic missiles on the open market, this regime forces countries like North Korea and Iran to develop their own missile technology.[32] The Chemical Weapons Convention restricts the transfer of equipment and chemicals necessary to the development of chemical weapons.[33] Not every attempt to institute international legal controls on weapons sales has succeeded, however. International efforts to regulate the trade in small arms have run aground on the opposition of the United States and a few other important countries, but nevertheless represent an important legal effort to limit the transfer of military equipment.[34] Most recently, various NGOs and advocacy groups have called for limitations on the export and use of autonomous weapons.[35]

In short, states (and defense firms within states) have a strong interest in exporting military equipment, but also face some important obstacles. Some of these obstacles take political form, while others rest on legal prohibitions. The arms trade, consequently, is ideal for investigating how intellectual property law might affect the diffusion of military technology across the international system. The rest of this chapter examines the various means by which states manage intellectual property concerns in the development, production, and diffusion of military technology. It concludes with a quantitative investigation of some aspects of the relationship between US arms exports and rates of intellectual property protection in buyer countries.

Licensing and Joint Production

Licensing represents the least complex form of intellectual property transfer. The licensing state receives technology and legal approval in return for financial and other considerations. Licensing agreements may also prohibit re-export of equipment, although only limited legal means exist to enforce this prohibition.[36] Historically, states licensed the production of military equipment outside their borders. Such agreements were common even in the communist world, which otherwise lacked strong intellectual property protections. In the wake of World War II, for example, the Soviet Union often licensed production of tanks, aircraft, and small arms to its satellites in Eastern Europe, as well as to the People's Republic of China. These agreements allowed Warsaw Pact countries to produce equipment such as the MiG-15, the AK-47, the "Whisky"-class submarine, and a wide array of other weapons under license.[37] This enabled the USSR to augment its own capabilities by increasing those of its allies and satellites, and to take advantage of and develop their industrial capacity.

The technology transfer associated with such deals could survive the diplomatic conditions that made the agreements possible. The People's Republic of China, for example, continued to produce Soviet-designed interceptors, submarines, and small arms long after the Sino-Soviet split.[38] The licensing state has limited tools for preventing such violations, beyond bilateral measures such as sanctions, tariffs, and the seizure of property. The expansion of the international IPP regime may offer more tools for enforcement, however.

Technology Transfer

The concept of "technology transfer" has merited a rich and extensive literature in the economic and development fields. Broadly speaking, scholars define technology transfer as the passing of industrial techniques, engineering know-how, and data from one country to another.[39] This necessarily requires the transfer of patents and trade secrets. Since at least World War II, technology transfer has played an important role in the arms trade. The devastating effect of the war on the industry of continental Europe and Japan, combined with the heavy American capital investment in military technology during World War II, led to a situation in which the United States enjoyed a massive technological advantage over its NATO partners and its non-NATO allies, an advantage that the United States became aware of during its postwar investigation of German industry.[40] Because the United States sought to increase European military capability as well as European economic prosperity, it facilitated the transfer of military and nonmilitary technology to European industry.[41] It placed stark limits on how its allies could use this technology, however, in order to prevent its transfer to the communist bloc (these practices are detailed in the section on export control below).

Contemporary arms negotiations have increasingly revolved around questions of technology transfer. For example, a recent deal between France and India involved the sale of thirty-six Rafale fighter aircraft, built in France and sold in flyaway condition.[42] This deal represented a huge victory for the French firm Dassault, which can now expect years of maintenance and supply contracts to keep the aircraft in operational condition. The Rafale deal emerged from the ruins of a much larger negotiation, which could have seen the joint construction of 126 Rafales between India and France, accompanied by the transfer of production technology that would have made India capable of managing the maintenance requirements of the Rafales on its own.[43] The agreement collapsed on the Indian requirement for joint production with Hindustan Aeronautics Limited (HAL), and on French concerns over the capacity of Indian industry to successfully manage production requirements.[44]

Originally considered a subcategory of technology transfer in weapons contracts, data transfer also represents an increasingly large concern in the arms trade.[45] In 2015, Japan agreed to the transfer of confidential data associated with its Soryu-class submarines to Australia, in an effort to capture a deal to build eight subs.[46] The data gave the Australian military a better understanding of the capabilities of the submarines, as well as the ability to successfully modify them, but it also left both the Japanese Maritime Self-Defense Force and Japanese shipbuilders vulnerable. As mentioned above, however, this is the classic tradeoff associated with international military technology transfer: increased cooperation and profit in exchange for potentially decreased national security. That Australia decided on a French bid instead of the Japanese bid highlights the risk of such transfers.[47]

Globalization of the Defense Industry

Accounts differ as to whether the drive toward the globalization of industrial production over the past few decades represents a break with the past, or simply the culmination of trends developing since the early modern era.[48] Almost all agree, however, that industrial production now has a more global character than ever before, with the components of finished products (cars, aircraft, computers) coming from an array of multinational corporate producers. Industrial firms have integrated across borders both horizontally and vertically, and have enthusiastically engaged subcontractors from around the world. Modern industrial production rarely happens in an exclusively domestic context.

For most of the twentieth century, the defense industry defied these trends. National governments jealously protected their defense firms, whether private or state-owned. They saw these firms not only as critical to national security, but also as potential vehicles for driving technological innovation.[49] After the Cold War ended, the number of firms focusing on the defense market contracted dramatically, but national governments retained an interest in protecting their defense industry, and in maintaining barriers to transnational collaboration.[50] At the same time, domestic law in many countries, including the United States, requires the production of some percentage of defense products domestically. Indeed, because of formal requirements associated with selling weapons to the United States, many foreign firms need subcontractor status in order to have any chance of taking advantage of the lucrative US defense market.[51]

The first steps towards breaking down these walls came in Europe, in the latter half of the Cold War. Although several of the biggest European countries maintained large defense industries (France, Italy, Germany, and the United Kingdom in particular), policy makers recognized the benefits of transnational industrial cooperation.[52] The increasing cost of advanced weapons, combined with the decline of global defense expenditures at the end of the Cold War, threatened to push many states out of the defense export market. Under the aegis of the European Union, however, several countries took steps towards defense industrial cooperation. Despite the EU's organizational capacity for fostering such cooperation, however, the major contributors maintained an interest in defending their own "national champion" defense firms.[53] For example, the main competitors for generation 4.5 fighter aircraft, a category that includes most modern avionics innovations but excludes advanced stealth, include Sweden's Saab Gripen, the Eurofighter Typhoon (sourced primarily from Germany and the United Kingdom), and France's Dassault Rafale.

Because the United States spends far more on defense than any other country, US defense technology companies enjoy several competitive advantages on the international market.[54] The flow of capital from the US government to the big US defense firms allows them to invest in nearly every corner of the defense market, while other countries need to specialize. Nevertheless, not even US defense contractors are immune to trends driving them towards globalization.[55] As discussed above, three trends appear to have moved the DIB inexorably towards globalization: the switch to becoming systems aggregators, the incentive to find the most competitive subcontractors, whether domestic or foreign; and the need to engage in purchases of dual-use technologies. As a result, transnational subcontracting partnerships have become common.[56]

Globalization of the defense industry has ongoing implications for the arms trade and for intellectual property protection. First, any relationship between a contractor and a subcontractor in a high-technology field requires agreement on the ownership of intellectual property. Subcontractors want to protect their trade secrets and patent rights, while the contractors want access to as much data as possible in order to ensure long-term viability. When the relationship between two firms crosses a border, it also crosses two systems of intellectual property protection, making the IP negotiation even more difficult.

Second, a firm's decision to subcontract across a border often requires the transfer of technology from the contractor to the subcontractor. This transfer can run afoul of the interest of national governments to maintain control of their defense technology, manifested in the form of export controls. This international technology transfer, again, requires a negotiation between two systems of intellectual property, along with management of the export control requirements and governmental demands for national security.

Third, the joint production, development, and subcontracting of dual-use components falls under the aegis of the international intellectual property regime, rather than under any arms-length negotiations between firms and governments. This means that even if firms and national governments try to avoid the procedures and requirements mandated by multinational intellectual property agreements, the subcontractors of their components still rely on those agreements in order to protect their property. Indeed, governments can sometimes use international subcontracting to short-circuit patent claims by domestic rights-holders. In *Zoltek vs. United States*, for example, a court held that while a government contractor had indeed infringed upon one of Zoltek's patents—part of the stealth covering of the F-22 Raptor fighter—the government was not liable, because the contractor had completed part of the process *outside* the United States.[57]

These complications can have significant geopolitical consequences. For example, concerns over intellectual property regulation have caused problems in the nascent military relationship between the United States and India. The growing power of China has concerned both India and the United States, making a deeper strategic partnership possible. Moreover, India has expressed considerable interest in several critical US technologies, including the electro-magnetic aircraft launch system (EMALS) that the US has developed for its most recent carriers.[58] However, the US defense industry remains concerned about cooperation with its Indian counterpart. Indian government rules prohibit foreign direct investment in its defense industry beyond 49 percent, which American companies fear would leave India in control of key technologies.[59] More broadly, India has resisted reform of many aspects of its domestic system of intellectual property, and has heavily criticized how the emerging international IP regime treats developing countries.[60] The US and India continue to work on reforming Indian IP protection procedures, but much work remains.[61]

Despite these complications, scholarly work has shown that the globalization of production in the defense sphere increases innovation; states taking advantage of globalization produce more innovative, effective weaponry, while those excluded fall behind.[62] The question becomes the degree to which states can pursue a strategy of globalization while also undertaking to protect their most innovative technologies from potential competitors. And, the following section on export controls suggests, the globalization of the arms industry offers important tools for managing the diffusion of military technology.

3D Printing

The development of 3D printing, which allows the production of three-dimensional solids without the need for an assembly line and traditional tool, die, and stamp production, could have a big effect not only on how military organizations maintain themselves in the field, but also on how they make long-term procurement decisions. Over the past several years the US military has experimented with 3D printing to manufacture components on military bases and even in underway naval vessels.[63] This reduces the logistical footprint of far-flung naval bases and warships, allowing frontline operators to manufacture intricate components and spare parts rather than rely on shipments from home. A 2015 report on China indicates that the People's Liberation Army Navy has equipped some of its deployed vessels with similar printing equipment.[64]

The conceptual foundations for 3D printing were laid in the 1970s, with the idea that machines could use lasers to render solid objects from a liquid mass.[65] Modern 3D printing uses digital design files to give the printer a template for the production of solid objects from preexisting materials.[66] These design files can be printed or transferred from a central source or can be created through scanning a physical object; a smartphone with the appropriate apps can in many cases create the necessary files.[67] Contemporary 3D printing can create a bewildering array of objects, from simple solids to complex parts to sophisticated combinations of solids and biological tissue.[68]

The popularization of 3D printing poses a threat to traditional manufacturing in that consumers may be able to avoid long lead time and investment in major manufacturing capacity, as well as the costs of transport of specialized materials and equipment. Widespread adoption of 3D printing also poses a significant threat to traditional intellectual property law. If 3D printing technology allows manufacturers to produce patented or copyrighted items without a large manufacturing base, then it limits the effectiveness of legal recourse against infringers. If property rights cannot be enforced, they lose value for their holders. The nightmare scenario for traditional IP rights holders involves an array of mobile 3D printers, each able to download design templates for medical devices, automobile parts, action figures, and other items that carry much of their value in intellectual property.

The advent of 3D printing also represents both a problem and an opportunity for states concerned about their military-oriented intellectual property. First, the problem: If weapons customers can create components and spares with advanced 3D printers, then they become less reliant on the MIC of the exporting country. Consequently, 3D printing gives manufacturers the technical ability to violate patent or copyright, thereby preventing producers from enjoying monopoly technology rights, and from controlling images and ideas. Manufacturers using 3D printers can also potentially skip over traditional trade secret protections, especially if the infringer can acquire design templates from illicit sources. Instead of buying new engines from Russia or France every time equipment breaks down, for example, Egypt could manufacture and assemble the most important components on its own.[69] Producing an air-to-air missile or an assault rifle won't become quite as easy to producing illicit copies of the film *Lord of War*, but it won't be far off.

The specialized manufacturing complexes that service a system of worldwide exports could become obsolete; and arms buyers could, if they so desired, simply ignore the intellectual property rights of the sellers. After all, why engage in complicated, potentially risky arms sales when you can just print copies of guns you bought originally? Traditional producers could lose the advantage they enjoy in specialized, expert workforces, such as those that exist in the aviation and shipbuilding industries. The problem gets even worse if consumers illicitly acquire the data and technical information necessary to produced specialized components and either sell it to a third party or use it to manufacture their own technology. This particular type of theft could severely undercut follow-on maintenance and service contracts, one of the pillars of the modern arms trade. In such a world, the protection of trade secrets becomes even more important, increasing the stakes of cyber security and industrial espionage conflicts.[70] In short, industrial espionage in the defense sector might get much more lucrative, and potentially a lot easier.

But the extent of this tragedy, such as it is, depends on the degree to which military technology consumers respect the intellectual property of producers. If buyers have good reason (either through the iteration of arms-length transactions, or because of national and international legal commitments) to avoid infringing on the intellectual property of sellers, then IP rights become central to the arms export relationship. Indeed, other than the initial frames of the technology being bought and sold—whether they're ships, planes, or vehicles—the key aspect of the arms deal becomes the intellectual property rights necessary to servicing and maintaining the transferred equipment.

The basic IP response for a firm includes the acquisition of design patents on specific design templates, which allows the firm some control over the data necessary to produce the items. As Pierce and Schwarz, members of the law firm Venable LLP, note, "Rights holders have a few options to pursue when they discover infringement. One such option is to sue the direct infringer, another is to pursue the parties who have knowingly aided in direct infringement, and yet another is to do both."[71] In the international intellectual property regime that the United States has helped create, this potentially could give US and other Western arms producers the opportunity to launch legal action against offenders, both in government and in private industry.

For producers, Pierce and Schwarz offer three recommendations for combating the threat to IP protection represented by 3D printing: increase the rate of technological innovation to stay ahead of infringers, adapt a business model that engages 3D printing, and build multiple layers of IP protection.[72] Not all of these solutions are adaptable to the military market, though we may expect that the second strategy in particular will become part of the arms export model. The need to develop technology faster in order to stay ahead of infringers could support a further shift to small, flexible production firms (a problem discussed in chapters 1, 3, and 4). The utility of multiple layers and formats of legal protection depends on the ability of states to enforce IP rights in the military sphere, but it could become a productive strategy as IP law further colonizes military production. For example, military producers could increasingly use copyright (a form of IP that normally only protects creative, non-useful objects, but which also protects software and design templates) to protect aspects of military equipment.[73]

In theory, it is possible for potential defense technology consumers to use 3D printing to avoid the technical violation of IP law by relying on open-source designs that effectively mimicked the capabilities of protected systems. For example, if key components to a MiG-29 engine became available through online open-source software and design specifications, then a 3D producer could manufacture the components without specifically violating Russian IP rights.[74] For this reason, some argue that the key to intellectual property protection lies in developing acceptable norms of behavior, rather than legal restrictions.[75]

There is already evidence that militaries have used 3D printing to evade foreign export controls. After the 2013 Ukraine War, Russia cut off supply of parts and spares to many of the Eastern European members of NATO. Most of these countries still use Soviet kit, including upgraded versions of Soviet fighters such as the MiG-29. Instead of starting over and purchasing entirely new systems from other sellers, the former Warsaw Pact countries turned to fabrication. Poland in particular has begun using 3D printing technology to replicate aircraft parts, relieving itself of the necessity of dealing with Russia in order to maintain the readiness of its fighter force.[76]

Export Controls

National security concerns deter some states from exporting their most advanced military and civilian technology. Officially and unofficially, many arms exporters employ export controls to prevent technology from ending up in foreign weapons systems. The desire to control the rate of diffusion of technology is the central reason why states place legal limits on weapons and dual-use technology transfers, but political factors also play a part in some export control systems.

This desire is especially strong in the US MIC. Export controls in the United States emerged as an effort by the state to limit the ability of private firms to send technology and finished weapon systems to potentially hostile foreign governments.[77] In the years immediately prior to World War II, the US government found that it had very few tools for managing the spread of advanced military technology, especially to Nazi Germany, Stalinist Russia, and imperial Japan. The case *United States vs. Curtiss Wright*, ruling that the Roosevelt administration had the inherent authority to prevent the export of military technology to Bolivia, offered the basic legal foundation for export management.[78] After the war ended, US planners believed they would require a significant technological advantage to offset the numerical superiority of the Soviet military, and they consequently helped institute strict rules on the export of equipment with military application.[79] For example, these rules forced private US firms to seek approval from the US government for the transfer of sensitive technologies. Essentially, the new regime placed limits on the rights of defense and other firms to control over their intellectual property.[80]

These rules established by the US government only became stricter during and after World War II. Through the medium of export controls, therefore, intellectual property protection formed a core aspect of US strategy in the Cold War. Moreover, the US strategy for technology management had an international aspect: although the US designed the system to prevent its own companies from transferring technology to the Soviet Union, in practice many friendly states found themselves the target of the export controls, due to concerns over secondary transfer and espionage. The international manifestation of export controls was the Coordinating Committee for Multilateral Export Controls, more commonly known as CoCom. Designed to coordinate high technology export policies across the United States, Western Europe, and Japan, CoCom came into effect in 1950.[81] The United States leaned hard on allied states, mostly Japan and the members of the NATO alliance, to limit the transfer of military and dual-use technology to the Soviet bloc, and to customers sympathetic with the Soviet bloc.[82] It sought to limit the export not only of technology that resulted from US transfers, but also of domestically developed technology that might enhance the capabilities of Soviet systems.

The export control system had a deep impact on the movement of people and knowledge. The system of protection that concentrated on the movement of "things" in the 1940s and 1950s soon turned its attention to "people."[83] This manifested in not only in visa regulation applied to international scholars and engineers, but also to classificatory schemes designed to prevent suspect individuals from accessing critical knowledge. Even the spread of unclassified information became problematic, if it might lead naturally to the revelation of classified knowledge.[84] Soviet efforts to collect vast reams of Western scientific knowledge undoubtedly heightened US concerns.[85] These efforts necessarily reduced the scientific capacity of the United States and its allies, both by compartmentalizing information and by insulating Western scientific communities from foreign knowledge and expertise. However, US policy makers believed that controls designed to limit personal interaction with Soviet and Soviet-sympathizing scientists would have a more negative impact on the USSR.[86] Such controls remained in force until the end of the Cold War, and concerns about Chinese espionage have revitalized the idea of curtailing foreign scientific contacts.[87]

Later in the Cold War, the role of export controls in maintaining American technological supremacy came under debate. On the one hand, the community of scholars and policy makers associated with the Office of Net Assessment in the Department of Defense emphasized the need for the United States to stay ahead of the USSR in technology in order to "offset" Soviet numerical superiority.[88] On the other hand, détente provided the basis for a variety of social and scientific exchanges between the US and the USSR. When détente waned after the Soviet invasion of Afghanistan, advocates of tighter controls gained the upper hand.[89] Even tighter restrictions on scientific cooperation and the export of dual-use equipment ensued. Similar arguments proliferated as the commercial technology relationship with China increased, to the extent that export controls were viewed as an appropriate means for protecting US market share in the international arms trade.[90]

The result of this decades-long effort is that the US has some of the most elaborate and careful export controls in the world, a complexity necessitated by its role as both economic and strategic global hegemon. Although the Directorate of Defense Trade Controls resides within the State Department, the responsibility for managing the system of US export controls lies across several government departments.[91] In the Department of Defense, the Defense Threat Reduction Agency plays a key role in evaluating technology transfer.[92] The Department of Commerce also plays a large role in the management of technology export.[93] In fact, US government export controls are so restrictive that prospective foreign buyers sometimes subject these organizations to lobbying efforts (as with the KF-X program, described in chapter 4), and also attempt to influence them indirectly through lobbying of Congress and other elements of the executive branch.[94]

The rules of export control are remarkably complex. Any equipment deemed to have military-only use falls under the responsibility of the Directorate of Defense Trade Controls (DDTC), requiring an export license. The DDTC compiles a large list of military-related items for regulation. States subject to US or UN arms embargoes (for example, China) cannot receive any of this equipment.[95] Even friendly states can run into difficulties if the DDTC deems technologies under consideration too sensitive for export; South Korea's inability to purchase several technologies related to stealth aircraft is discussed in chapter 4.

The restrictions imposed by export controls also apply to dual-use equipment. The United States prohibits the export of an array of dual-use technologies with the potential to improve the capabilities of foreign weapons. These prohibited systems include advanced computing technologies necessary for running simulations or for improving the precision of manufacture.[96] The spearhead of the US government dual-use technology export control system is the Department of Commerce, which manages dual-use equipment through the Bureau of Industry and Security. This bureau classifies technologies by type and sensitivity, regulating the export of equipment based on target country.[97] This classification is based on the potential national security threat posed by the technology consumer; US allies are subject to fewer restrictions than are neutral or hostile countries. Dual-use items subject to export control restriction run the gamut, with the most important items including high-tech electronics and sophisticated software.[98]

Complex and stringent as the US export control system is, it is not foolproof. The Toshiba-Kongsberg scandal was one of the most well known failures of export controls during the Cold War. The Japanese firm Toshiba and the Norwegian firm Kongsberg each operated under strict export controls designed to prevent the sale of advanced manufacturing and computing technology to the Soviet Union.[99] The United States suspected that such technology could significantly improve the design and manufacture of submarine propulsion systems, thus reducing the noise of Soviet submarines. In the early 1980s, Toshiba supplied milling machines to the Soviets, while Konigsberg, acting on Soviet design requests, supplied the software necessary to drive the machines. Both companies violated domestic export control regulations, in addition to incurring reprisal from the United States.[100]

Export controls do not generally cover single weapon systems in this fashion, but rather cover subsystems that contribute to the overall effectiveness of a weapon or class of weapons. Under special circumstances, however, export controls can apply to entire systems. Congressional concern over the transfer of F-16 technology from Israel to China led to the Obey Amendment, a law intended to prevent export of the F-22 to any foreign government.[101] Similarly, strategic arms limitation treaties prohibit either Russia or the United States from transferring nuclear-capable strategic bombers.

Export controls can generate tension between private firms and the government, especially for firms that normally focus on the civilian market.[102] This tension becomes even greater when firms globalize; a 2015 report in the *New York Times* detailed the many connections between US technology firms and their Chinese counterparts.[103] The report suggested that many companies that US firms regularly deal with have their own relationships with the Chinese military, and consequently may run afoul of export control rules in ways that are difficult to police. The *Times* report, inspired by a report from the private intelligence firm Blue Heron on IBM's dealings with China, highlighted the tensions between maintaining security over American dual-use technological innovations and staying abreast of the global technology market.[104] While the report does not indicate that IBM has violated the US system of export control, it does imply that the system lacks capacity to properly monitor interactions between US and Chinese companies.

A 2012 article in *China Business Review* detailed some warnings for private firms working in technology cooperation with China, given the existence of US export control rules.[105] These include

- allowing Chinese workers to manufacture sensitive US technology,
- conducting research and development that uses US software or technology,
- collaborating with Chinese researchers on nonpublic intellectual property,
- selling to Chinese consumers,
- supplying tech support to Chinese consumers,
- licensing software or technology to Chinese firms,
- engaging in corporate transactions with Chinese investors, and
- facilitating visits of Chinese nationals to US manufacturing or research facilities.

In spite of the obstacles that the US export control regime erects that are designed to prevent or limit US firms from collaborating with their Chinese counterparts, the United States and China nevertheless have established a massive collaborative economy, often in high-technology equipment. This suggests that some companies may honor many of these restrictions in the breach.

This collaboration is understood by US firms to pose an IP protection risk, however; and in response to concerns about IP enforcement in China, some firms have begun to develop more sophisticated strategies for protecting their property. The Taiwanese Semiconductor Manufacturing Corporation (TSMC), for example, recently opened a new plant in China, without a Chinese partner.[106] This makes the physical protection of trade secrets easier, and also simplifies the legal defense of TSMC's property. In addition, TSMC has maintained facilities for the production of its most sophisticated chips in Taiwan, thus preserving a degree of market advantage. US firms can avail themselves of similar strategies as long as the Chinese economy remains open, though facilities constructed in China will always be at greater risk of penetration than facilities built in the United States.

Some Implications of the Emerging International Intellectual Property Regime for the Global Defense Industry

As discussed above, the defense sector has experienced globalization differently from most other industrial sectors. Because of the desire of national governments to both protect technology and foster domestic firms, defense firms have retained a more national character than most other kinds of corporations. This national character has limited the exposure of the defense industry to the pressures that drove other multinational corporations to develop and promulgate international standards of IP protection. The defense industry was largely absent from the consortium that produced TRIPS, the most serious effort to create international intellectual property standards.[107] Consequently, the rules created and promulgated through the WTO and other organizations did not fit comfortably with the needs and interests of global defense industries.

As this chapter has detailed, states and firms have long had an interest in managing concerns over technology transfer and IP rights. The prospect that American military or dual-use technology might fall into the hands of the Soviet Union caused endless concern in Washington during the Cold War, even as the United States attempted to learn everything it could about Soviet technology and production methods.[108] For the most part, however, states dealt with intellectual property problems through arms-length transactions: licensing deals, specific technology transfer arrangements, and limited joint production and development projects. The export control regime developed by the United States but enforced multilaterally is the major exception to this trend.

The question that animates this chapter is the extent to which the transformation in how global industries have viewed and managed IP protection issues has extended or will extend to the defense industry. If the answer is "not much," part of the reason concerns the lack of a serious need for defense technology companies to seek this protection. Even if defense firms lack interest in doing so, however, producers of dual-use technology will undoubtedly find it necessary to protect their intellectual property abroad.

Since the end of the Cold War the United States and, to a lesser degree, Western Europe have moved into a dominant position within the international arms trade. This dominance is reflected not only in sales figures but in the extent to which US and European firms have diversified supply and subcontracting relationships across the world. As we would expect, the states most interested in imposing robust IP protection for high-technology products are those that dominate the defense export industry.

Conceivably, the defense industrial market could develop into a two-tiered system of military technology export and production, with the first tier including the United States, Europe, Japan, South Korea, and Australia, and the second tier including China, Russia, and Iran. Relations between first-tier exporters, and between first-tier exporters and their customers, would be characterized by a detailed, robust, multilateral system of intellectual property protection. Relations in the second tier would lack such exacting standards, but could include states that reject the first-tier IP protection system. Export procedures in this tier would continue to take IP protection into account, but mainly through arms-length contracting rather than through multilateral standards. States such as India and Brazil might float between the two tiers, depending on their specific military or dual-use technology needs. In this system, which can be thought of as the "anything goes" system of IP management, states beg, borrow, and steal whatever technology they can, often attempting to copy or reverse-engineer systems developed in other states. In many ways, Iran and North Korea operate by the same rules, buying and exporting missile technology without respect for legal protections on the patents or trade secrets that go into them. This older, more traditional system of "anything goes" IP management also allows countries like North Korea, Syria, Iran, and Myanmar to gain access to weapons without the acquiescence of Washington or the major European arms exporters.

The near-term development of an alternative system of IP protection among the second-tier states is possible, but not particularly likely. First, as has become obvious in the Russia-China relationship, arms-length arrangements do not suffice to give states the confidence to export military equipment and technology to potential rivals. Even if China and Russia do not view each other as strategic threats, they surely view each other's military industries as economic competitors. If they desire to further exploit arms markets in Latin America, Southeast Asia, and Africa, they will eventually need to sell to countries that may have the capacity to appropriate, reproduce, and then re-export their technology.

Second, any two-tier system that Russia and China create, formally or informally, will still require technology transfer from first-tier states. Every innovative industry in the world requires technological input from international sources.[109] Neither Russia nor China can operate on the frontier of military technology without access to the latest in dual-use systems, and as long as these systems are available only from companies in first-tier states, (cyber espionage efforts aside) they will find it difficult to access such systems if they maintain a second-tier-level system of IP protection.

While neither China nor Russia may have a strong commitment to the maintenance of a robust international IP protection regime, both may in time benefit from adherence to international standards.[110] For example, China's position in the military export market and in the larger international economy is growing more significant. As it comes to rely on an export market for military hardware, the salience of IP concerns will grow.[111] By and large, exporters of high-technology equipment benefit from the strict enforcement of IP rules. As Chinese military production increases in sophistication, and as the Chinese military supply chain spreads across different countries, China's position on IP management for military equipment may become more similar to those of first-tier states than to those of states in the second tier. Like the United States, China may always have political reasons to export weapon systems to bad IP citizens. In the future, however, it may pay increasing attention to precisely how its technology is used, and it may attach strings intended to prevent the loss of Chinese trade advantages. In this sense, the politics of intellectual property are just like any other kind of power play: states seek to structure the rules of the game to their own advantage. As the economic and military profiles—if not necessarily the interests—of the United States and China converge, there will be more substantial grounds for cooperation on the rules that govern the international IP regime.

Table 5.2 depicts the world's major arms producers and the average intellectual property protection scores of their customers. Trend indicator value comes from SIPRI, while average IPP scores come from the World Bank database. Ranking by tiers, common practice in the arms trade literature, is based on technological sophistication of the DIB.

Finally, if either China or Russia wants a forward-looking technology sector with a strong collection of private firms, they will need to strengthen their domestic IP protection regimes. As chapter 3 noted, China has already begun this process. Both China and Russia will find it exceedingly difficult to put into place a system of intellectual property protection that fosters an

TABLE 5.2. Average intellectual property protection scores for arms sales producers, 2012–16

Tier/region	Country	IPP average	TIV total
Tier 1			
Europe	France	5.71	8,579
	Germany	5.55	7,663
	Italy	3.78	3,678
	Russia	2.80	33,857
	UK	5.88	6,637
	Europe average	4.74	12,083
Americas	United States	5.27	46,790
Tier 2			
Europe	Netherlands	5.83	2,797
	Spain	3.93	3,994
	Sweden	5.67	1,648
	Europe average	5.14	2,813
Middle East	Israel	4.73	3,446
Asia	Australia	5.45	417
	China	3.96	8,690
	India	3.76	133
	Japan	5.68	N/A
	South Korea	4.06	1,415
	Asia average	4.58	2,131
Americas	Brazil	3.44	267

environment necessary for domestic firms to flourish, but which leaves foreign firms without recourse to IP infringement.

Quantitative Investigation

We have conducted a limited statistical investigation in support of the argument that the intellectual property considerations have some impact on US decisions on arms exports. We analyzed international arms sales made by the United States to other countries from the years 2012 to 2016. We gathered this information using the Stockholm International Peace Research Institute's (SIPRI) arms transfer database, which uses trend-indicator value (TIV) to operationalize the value of each arms deal. We used this TIV for each country-year as our dependent variable.

To explain how the TIV from the United States to client countries might vary, we conducted a regression analysis using the following independent variables: the World Bank's IPP ratings, where each country's intellectual property protection is rated from 1 (worst) to 7 (best); polity IV scores, from -10 (worst) to +10 (best); a dichotomous variable "conflict," measuring whether a client country was involved in a conflict that year, according to the Uppsala Conflict Data Program's definition of a "conflict" involving at least twenty-five battle deaths per country-year; a dichotomous variable measuring whether the client country was a signatory of the TRIPS Agreement, administered by the WTO; an interaction variable multiplying IPP and polity; and five financial control variables based on World Bank data, including military expenditures as a percentage of GDP, personal purchasing power logged and unlogged, and GDP logged (table 5.3).

The regression analysis demonstrates that the relationship between IPP and TIV is both positive and significant, reinforcing the argument that the United States takes the client country's level of intellectual property protection into account when deciding to sell arms to that country. The IPP independent variable had a significance level of at least 0.05 in every model, and a significance level of at least 0.01 in four out of the six models, including the model containing just that variable. This relationship is driven in part by the United States' healthy arms sales relationship with countries like Saudi Arabia, Qatar, the United Arab Emirates, and Singapore, which place a great deal of importance on intellectual property protection, if not necessarily on democracy.

The presence of conflict also has a positive and significant relationship with TIV, although it is not as strong as that of IPP. While this relationship would seem to make a certain amount of sense—countries at war are more likely to buy weapons than countries not at war—the regression results suggest that the relationship is not ironclad. The relationship between conflict and TIV is significant in only three of the five models in which it is included, and at varying levels of significance. Granted, all three of those instances are positive, suggesting that countries involved in conflict—Turkey, Iraq, India, Pakistan,—are more likely to be US arms clients than those that are not.

Like IPP, membership in the WTO and being a signatory to the TRIPS agreement has a significant and positive relationship with TIV, suggesting that membership in an international trade organization and a public commitment to protecting intellectual property are more likely to make a country an attractive arms trading partner for the United States than not. Of the five models in which the TRIPS variable is included, it is significant in all of them, with a *p* value of at least 0.01 in three of them. In addition, the coefficients are all positive, suggesting that the relationship between WTO membership and US arms sales probability is a strong one.

This relationship, however, could also be the result of the client country's willingness to pay, which would help explain the positive and significant relationship between TIV and percentage of GDP devoted to military spend-

TABLE 5.3. Predicted probabilities for US arms transfer

	Model 1	Model 2	Model 3	Model 4	Model 5	Model 6
(Intercept)	−115.21***	−263.95***	−915.73***	−215.90***	−323.60***	−1100.57***
	(−26.91)	(43.49)	(99.56)	(48.82)	(52.14)	(103.51)
IPP	45.11***	42.22***	20.64**	21.62**	58.94***	66.01***
	(6.72)	(7.07)	(9.42)	(10.97)	(12.96)	(12.66)
Polity		−2.00	−2.70*	−2.61	26.13***	24.75***
		(1.60)	(1.63)	(1.61)	(5.82)	(5.50)
Conflict		43.18*	−0.04	44.30**	55.93***	5.85
		(21.99)	(22.66)	(21.83)	(21.47)	(22.16)
TRIPS		99.76***	94.57***	107.02***	82.20**	67.71**
		(35.77)	(34.34)	(35.51)	(35.05)	(33.94)
Milex/GDP		36.59***	38.81***	39.92***	30.21***	29.43***
		(4.63)	(4.72)	(4.74)	(5.01)	(4.95)
Log (Ppp)			8.32			8.93
			(8.46)			(8.27)
Log (GDP)			24.25***			26.21***
			(7.28)			(7.12)
GDP per capita				0.00**	0.00***	
				(0.00)	(0.00)	
IPP × polity					−7.53***	−7.09***
					(1.47)	(1.36)
R^2	0.07	0.22	0.30	0.24	0.27	0.33
Adj. R^2	0.07	0.21	0.29	0.23	0.26	0.32
Number of observations	622	555	554	554	554	554
RMSE	182.25	169.95	162.04	168.37	164.60	158.29

***$p < .01$
**$p < .05$
*$p < .01$

Sources: World Economic Forum Global Information Technology Report 2016 (https://www.weforum.org/reports/the-global-information-technology-report-2016),Center for Systemic Peace Polity IV Research Project 2017 (http://www.systemicpeace.org/inscrdata.html), Uppsala Conflict Data Program (http://ucdp.uu.se), World Trade Organization (https://www.wto.org/trips), and World Bank (https://data.worldbank.org).

ing. Again, the logic here seems straightforward: countries that spend more money on their militaries are going to buy more arms, and the relationship between the two variables might be robust enough for this to be the case: the military expenditure variable is positive and has a *p* value of less than 0.01 for all five of the models in which it is included. These results imply that the United States' best customers for arms tend to be the most enthusiastic purchasers of arms overall, even to the exclusion of other types of government spending.

The relationships between the variables paint a relatively clear picture of the type of state to whom the United States is most likely to sell arms. That state is likely to be a member of the "community of nations" participating in international trade agreements like TRIPS, and will protect its corporate citizens and the countries with which it does business through vigorous intellectual property protection. Occasionally, this protection will extend to the rights of its citizens who are not corporate, and this IPP and international cooperation will be a result of its commitment to democracy and good global citizenship.

But this commitment to protection and good citizenship is not always the case. Sometimes the state must focus on arms sales because it is experiencing conflict and buying arms as a result. Given that most of the United States' most enthusiastic customers are in conflict-prone regions—nine of its top ten customers are in Asia or the Middle East—this conclusion does not seem unreasonable. However, given the relatively weak support for it, combined with the fact that two of these top twenty are Iraq and Afghanistan—both in a region prone to conflict partly as a result of US actions—this is a conclusion we must be careful about drawing.

Instead, it is important to observe that the variables with the most consistent and abundant support—IPP, TRIPS, and military expenditures—paint a very specific picture of the type of country to whom the United States is most likely to sell arms: one that cares very much about the rights of businesses, and which is willing to spend a great deal of its budget on weapons whether it is in conflict or not. Of the top five US customers—Saudi Arabia, the UAE, Turkey, Australia, and Taiwan—only Turkey was involved in any sort of conflict. Reinforcement of democracy and assistance to conflict-torn countries in need are not very reliable drivers of the US decision to sell arms; countries that have put IP protections in place and are willing to buy are more reliable candidates, regardless of their politics or motivations beyond a desire to participate in the global marketplace.

Conclusion

The arms export industry has come late to an understanding of the importance of IP law. Historically, political and economic factors have posed barriers to the theft of defense-technology-related IP. States and firms have treated IP protection as an arms-length concern, managed through individual contracts and relationships rather than through an overarching framework. The primary systematic means of managing technology transfer came as a result of concerns over the export of US military technology to the Soviet Union and its allies. This system of export control developed into a de facto international intellectual property protection regime during the Cold War, and it continues to regulate the transfer of military technology today.

However, changes in the nature of defense technology, in the structure of the defense industry, in the nature of manufacturing, and in the construction of international intellectual property law have made defense exporters, both in private firms and in governments, more cognizant of IP concerns. Dual-use technologies require a more systematic and careful application of IP legal principles than do legacy military equipment, as private civilian-oriented firms have less tolerance for export restrictions than do their defense-oriented cousins. The development of the defense firm into an aggregator with an international system of subsidiaries and subcontractors has made managing the transfer of military technology more complex, both in terms of various cross-border arrangements and in the negotiation of agreements between private firms.

Finally, while international and domestic legal considerations have long constrained the ability of states and firms to export weapons solely on the basis of demand, the emerging international IP protection regime may soon exert its own effects. The ability of states and firms to use this regime to protect their IP internationally may change how states purchase, maintain, and re-export military technology. Specifically, the international availability of legal sanctions that have heretofore only applied to the domestic setting may make states more willing to allow the export of advanced, protected technologies.

Some important questions remain for further research: Has the expansion of international intellectual property protection changed how states export weapons? Has the expansion of international intellectual property protection changed how states develop weapons? Do states curtail weapons exports because of concerns about insufficient IP protection? Do states curtail joint production, licensing, and subcontracting arrangements because of concerns about insufficient IP protection?

6: INTELLECTUAL PROPERTY, INDUSTRIAL ESPIONAGE, AND CYBER SECURITY

Vignette: Stealing a Bomber

In 1944, during the course of the American strategic bombing campaign against Japan, five Boeing B-29 Superfortress heavy bombers unexpectedly landed in the Soviet Union.[1] While the United States and the USSR were allies in the European theater of operations, in the Pacific the Soviets remained neutral, maintaining correct diplomatic relations with Japan until mid-1945. Rather than returning the B-29s and several other aircraft that had landed in Siberia, the Soviets decided to intern them for the duration of the war.[2]

The Superfortress represented an enormous corporate investment on the part of Boeing, and national investment on the part of the US Army Air Force. The bomber, intended to fly in self-defending high-altitude formations, represented a key component of the Air Force's expectations for dominance in the postwar world.[3] It became the central aircraft in the Pacific bombing campaign, with a pair of Superfortresses eventually delivering the atomic strikes on Hiroshima and Nagasaki. By some accounts, the B-29 program cost the United States more than the Manhattan Project.[4]

The Soviet Air Forces, despite their reputation as a ground support force, had invested heavily in strategic bombing prior to World War II.[5] Unfortunately, none of their projects had quite panned out, and the Soviet High Command frittered away most of the bombers in a fruitless attempt to stop the German advance in the early days of Operation Barbarossa.[6] The Soviets requested access to the B-29 through the lend-lease program, but the United States had little interest in giving one of its most impressive technological achievements to an ally of uncertain loyalty and intentions.[7]

But with the four intact and one damaged bombers now in their possession, the Soviets had the opportunity to closely study the aircraft, and perhaps attempt to replicate it. Although the Soviets generally dismissed the effectiveness of the strategic bombing campaigns against Germany and Japan, they saw some value in having a heavy strategic bomber of their own, especially if reverse engineering could spare them the development costs. As a bonus, several of the aircraft carried technical manuals detailing the maintenance and performance of the bomber.[8]

Stalin ordered the USSR's best aviation engineers to copy the B-29, and ready it for serial production. The aircraft included a variety of features utterly unfamiliar to the Soviets, including "lightweight aluminum alloys, pressurized crew compartments, remote-controlled guns, powerful super-charged engines, Norden bombsight, radar, electronics, and instrumentation."[9] While some Soviet engineers argued for incorporating this technology into existing Soviet designs (the B-29 was already growing long in the tooth by the standards of World War II aviation), Stalin insisted on a copy.[10]

Within about two and a half years, the Soviets had a prototype, the Tupolev Design Bureau's Tu-4. By 1949, they began to deploy the Tu-4 in numbers. Not everything on the Tu-4 was illegally appropriated; the Soviets had acquired licensing rights to some US engines during the war, and they installed modified versions in their own bombers.[11] Despite having several extant examples, which allowed both full disassembly and comparison to a complete example, the process of reproduction of the Superfortress hit a number of difficult snags. Soviet tool systems, based on metric units, were not compatible with the systems used in US industry, which used English measurements; and this forced a number of minor modifications. The effort proved demanding for Soviet industry, eventually occupying sixty-four design bureaus and some nine hundred factories.[12]

Nevertheless, the Tu-4 flew in Soviet Long Range Aviation for almost two decades, with 847 aircraft eventually entering service. The Soviets even exported ten Tu-4s to the People's Republic of China, with the last aircraft leaving service in 1988.[13] After the Sino-Soviet split, China would show the same degree of courtesy to Russian equipment that the Soviets showed to the United States, reverse engineering or otherwise copying a variety of systems, including the MiG-21 fighter. The availability of the B-29 also helped jump-start the Soviet aviation industry, allowing it to catch up with the United States far more quickly than it otherwise would have done.[14]

What does this have to do with cyber warfare? Whether undertaken by states or private actors, and whether involving patents or trade secrets, industrial espionage in general represents the theft of intellectual property. Cyber warfare is a means of industrial espionage, which continues to play an important role in the diffusion of military technology. During the Cold War, both the West and the Soviet bloc delighted in the acquisition of each other's aircraft, tanks, and other weapons. New equipment was disassembled, subjected to an array of tests, and used to modify industrial priorities. While some of these practices continue (China's acquisition of fragments of a downed F-117 stealth fighter-bomber in 1999 is a modern example), the digitization and interconnection of industrial knowledge has largely made these techniques of espionage secondary.[15] Today, industrial espionage has taken a different form. By most accounts, over the past decade and a half China has devoted significant resources to efforts to steal digitized US and European intellectual property through cyber attacks. These attacks have targeted both civilian and military facilities, stealing technologies with civilian, military, and dual-use application.

Introduction

Since the early days of the industrial revolution, national governments have sought to steal the economic and manufacturing secrets of foreign countries. This theft has ranged from the passive (lax or no enforcement of protection on foreign IP) to the active (the dispatch of agents directly to foreign countries in order to observe and mimic processes).[16] In the eighteenth century, for example, the French government sent agents to industrial centers across Great Britain in an effort to both steal processes and potentially lure expert workers across the English Channel.[17] Given that national power depended on the sophistication and productivity of industry, and that major national stakeholders sought to improve their position on the global market as a way of maximizing both profit and national security, the existence and range of this espionage should surprise no one.

Indeed, as chapter 2 indicates, national authorities have only recently come around to the idea that respect for IP should cross international borders. The motivating concept for IP protection is that legal defense for inventors can allow them to retain a temporary monopoly over production of their invention. The granting of this temporary monopoly provides incentive for innovation, benefiting the entire public. In the nineteenth century, "public" applied almost exclusively to the domestic context, to the extent that inventors who managed to copy or appropriate foreign intellectual property were regarded as national heroes. The expansion of "public" to refer to an international community of property holders has happened only recently, at considerable political expense, and with great controversy.[18]

During the Cold War, concerns over industrial espionage loomed large. The key military systems of the Cold War relied on sophisticated components that spying could steal or make vulnerable. The development of jet, missile, nuclear, submarine, and radar technology meant that either the United States or the Soviet Union could potentially reap huge advantages from intricate knowledge of its foes' systems. These concerns prompted the

US government to develop the extensive system of export control discussed in chapter 5.

The United States and the Soviet Union waged an asymmetric war over industrial espionage. Soviet technology consistently lagged behind that of the West, though Soviet industry offset US competitive advantages in terms of cost and productivity.[19] In order to remedy this lag, the Soviets often sought to acquire Western technology through illicit means, either to directly copy foreign systems or to introduce new technological processes and know-how into the Soviet system of national innovation. And the Soviets were occasionally successful. Observers of the military aviation industry, for example, could hardly fail to notice the tight similarity between generations of US and Soviet bombers, from the aforementioned B-29 and Tu-4 to the B-52 and Tu-95, the XB-70 and T-4, and the B-1B and Tu-160.[20] While some of these similarities were the result of parallel development, there is direct evidence that direct appropriation of US technology affected Soviet bomber development.[21]

For its part, the United States also successfully stole Soviet industrial secrets. However, it only rarely sought to acquire and copy complete systems.[22] Rather, US espionage efforts focused on learning as much as possible about foreign systems, giving domestic industry the opportunity to develop countermeasures. For example, Adolf Tolkachev, a Russian radar engineer, spied for the United States over the course of six years in the late Cold War.[23] Driven by a disdain for the Soviet system, Tolkachev turned over a vast amount of information regarding Soviet airframes, Soviet electronics, and Soviet radar systems. This intelligence supplied the United States with a key strategic advantage—it made clear, for example, that the Soviets had no good answer for US systems such as ground-launched cruise missiles—and also critical data on the effectiveness of Soviet electronics. Working in conjunction with private industry, the US government did not directly appropriate Soviet systems, but instead used the information to alter the trajectory of America's own aerospace and electronics efforts.[24]

In short, the United States used Tolkachev's trove to pursue strategic advantage against the Soviets, but not to directly appropriate Soviet breakthroughs. US intelligence repeated this pattern in other areas, including the acquisition of data on Soviet fighter aircraft, missiles, and submarines. Over time, US officials have stressed the distinction between the use of industrial espionage to gain strategic advantage (which they consider legitimate), and its use to directly appropriate foreign secrets (which they consider illegitimate).[25] Nevertheless, this kind of industrial espionage paid enormous benefits in terms of American technological and strategic planning during the Cold War.

The question animating this chapter is whether the combination of changes in technology change and changes in the practice of intellectual property protection have introduced a new era in industrial espionage. We argue yes; industrial espionage now largely takes the form of cyber attacks against the IP holdings of state, military, private, and legal institutions. The next section investigates the dynamics, history, and processes of a specific form of industrial espionage commonly known as reverse engineering. Following this, we discuss the emergence of cyber security, cyber war, and cyber espionage both as the most recent evolution of industrial espionage and as national security concerns. The chapter then details the US-China cyber relationship, with particular attention not only to espionage but to efforts on both sides to create norms of behavior that facilitate cyber restraint.

Reverse Engineering

As mentioned in the introduction, one means of violating intellectual property rights in the military sphere is the reverse engineering of technology acquired through legitimate or illegitimate means. Reverse engineering involves taking apart an example of an extant piece of equipment, and then attempting to reproduce it.[26] As a practical matter, reverse engineering of foreign technology for domestic production faces several difficult obstacles: a lack of trade secrets, a lack of testing data, and the lack of a broader industrial ecology.

The obstacle of the technological mismatch is, at its root, that the thief lacks trade secrets associated with the manufacturing of the system. At the very least, this absence can make the replication of foreign systems a costly and time-consuming process, as the thief needs to develop manufacturing procedures from scratch. At worst, it can lead to seriously substandard components that reduce the capabilities and reliability of a system. For example, Chinese efforts to reverse engineer certain Russian jet engines during the 1990s and 2000s invariably produced engines with extremely short lifespans, and without the power of their Russian counterparts.[27]

Second, the thief lacks access to data associated with design and testing. Modern weapon systems generate an extraordinary amount of data during the development process, as computer models explore a vast array of scenarios with respect to potential components.[28] The testing process also generates data, and the thief generally lacks access to prototype models and to testing data associated with the system. This makes it difficult to come to solid conclusions regarding the tolerances of particular materials, or even the purpose of certain subcomponents, not to mention the overall reliability and performance capabilities of the technology in question. In the case of the USSR's theft of the B-29, the Soviets needed to conduct extensive materials testing of their own in order to mimic American manufacturing.

Third, the success of reverse engineering depends on access to a broader ecosystem of technologies. As Andrea and Mauro Gilli have pointed out, successful adoption of a given military technology often requires adoption of a broader family of associated technologies. An armed drone, for example, requires broadband, communications, and surveillance technologies that extend far beyond the airframe itself. In this sense, acquiring a unit for reverse engineering, or acquiring blueprints through cyber espionage, does not necessarily produce meaningful diffusion.[29]

Altogether, reverse engineering an entire military system is generally more trouble than it is worth. States that have the industrial and technological capability to reverse engineer a complex system generally also have the capacity to engage in their own design work.[30] Domestic designs mean that the builder can focus on the weapon characteristics it wants, rather than settling on a system designed by a foreign producer. Nevertheless, as the example of the B-29 and the Tu-4 shows, reverse engineering occasionally works in spite of the odds. Some states, most notably China, have persisted in efforts to reverse engineer foreign military technology.[31]

Allied efforts to reverse engineer German technology in the aftermath of World War II provide a real-world test case of the utility of reverse engineering. Shortly after the war, the United States, France, the United Kingdom, and the Soviet Union all attempted to appropriate German intellectual property in pursuit of "intellectual reparations," technological and economic repayment for damage caused during World War II. Most of these efforts encountered difficulty turning static intellectual property—patents, diagrams, formulas—into technological innovations, at least until they began directly borrowing German scientists and engineers. Tacit knowledge (know-how) was critical to the process of reverse engineering.[32]

Industrial Espionage and the Digitization of Knowledge

Over the summer of 2015, the popular national security blog Lawfare asked readers to name the most hackable database operated by the US government. The request came in the wake of a massive hack of the Office of Personnel Management database in spring 2015, which led to the loss of huge amounts of personnel information on government employees and applicants. The primary authors and readers of Lawfare submitted dozens of databases, most of which contained various types of information about government personnel.[33] Two of the databases, however, made a different kind of sense. One, cited by Ben Wittes, was the Commerce Department's database of export control applications, which determine how and where US firms can export sensitive dual-use technology.[34] Reader Jonathan Lichtman suggested the US Patent and Trademark Office database, which includes applications, with supporting materials, for US patent protection.[35]

Lichtman's idea made sense, because the concurrent development of cyberspace and the expansion of intellectual property law have changed the context in which states conduct industrial espionage.[36] The digitization of knowledge means that patent applications, trade secrets, and reams of industrial data have become available to talented hackers and dedicated organizations. Moreover, the functioning of intellectual property law in the United States and elsewhere requires a degree of communication between different organizations. The military services, contractors, subcontractors, defense firms, and law firms all have some degree of access to crucial secrets. Hackers can attack these communications, in addition to the home organizations. For example, the Mandiant: APT1 report (a 2013 report by the cyber security firm Mandiant) indicates that People's Liberation Army hackers stole information directly from US law firms specializing in intellectual property.[37]

In a sense, the vulnerability arises from changes in the nature of the defense industrial base, and more broadly in the nature of modern capitalism. Specialization of firms increases interfirm communications, which then results in communications vulnerabilities: the more firms need each other, the more vulnerable they become to communications hacks. Cooperation with the regulatory state complicates the picture even further. Many companies and legal firms have already begun to take steps to manage their vulnerability, including developing firewalls on communication with foreign clients and affiliates, especially those in China.[38] However, hackers have the luxury of concentrating on the weakest links. They can probe every member of the prime contractor-contractor-subcontractor-government agency relationship, looking for security vulnerabilities that allow them access to data, engineering files, and even other partners.

Scholarship on the development of the information economy has long grappled with the shift from an industrial to a postindustrial knowledge-based economy.[39] Modern computing technology has enabled the collection of tremendous amounts of data, with processors allowing for search and analysis, and communications equipment facilitating near-instantaneous transfer of information. Decades ago, the information contained in the databases mentioned above resided in huge warehouses, and could not effectively be "stolen" without immediate physical presence and the use of heavy equipment. For example, in 1969 Daniel Ellsberg photocopied some seven thousand pages of material associated with US government deliberation on the Vietnam War. Ellsberg smuggled these materials out of the RAND Corporation over the course of several weeks before passing the documents to the *New York Times*.[40] By contrast, in 2009 Chelsea Manning gained access to several classified databases, which she then leaked to the transparency advocacy network Wikileaks. The entire database (around five hundred thousand documents) fit on a single rewritable CD and was communicated to Wikileaks electronically.[41] In 2013, NSA contractor Edward Snowden leaked a very large—the exact number is unknown—number of documents to a group of journalists.[42] In both cases, releases of this scale would have been difficult or impossible if the leakers had lacked access to modern digital technology.

This digitization of knowledge extends to the intellectual property sphere. Paperwork associated with trade secrets once resided in guarded offices, behind physical lock and key; now it exists on computers with varying levels of protection from outside attackers. Copyrighted material such as film, music, and books can be digitized and shared almost infinitely; thieves no longer need to produce physical copies of pirated material.[43] Finally, while most patents (outside those designated as having national security significance) have always been in the public sphere, these patents are now far more accessible than in previous decades. Whereas access to these patents once required physical presence at the US Patent and Trademark Office in Washington, they now only require an Internet connection. This enables far broader dissemination of the knowledge embedded in those patents.

Cyberspace, Cyber Conflict, and Cyber Espionage

Cyberspace is a relatively recent development; the necessary connectivity only emerged in the 1970s, and the bulk of the public only became aware of the Internet's existence and utility in the 1990s.[44] The development and impact of cyberspace has attracted attention from a variety of disciplines, including political science, organizational theory, economics, business, computer science, sociology, and various technical fields.[45] It has also produced a huge literature in military, law enforcement, and public policy fields. Given the novelty of the cyberspace domain and the wide array of disciplines working within it, some confusion over terminology is inevitable. In the words of Brandon Valeriano and Ryan Maness, cyberspace constitutes

> the networked system of microprocessors, mainframes, and basic computers that act in digital space. Cyberspace has physical elements because these microprocessors, mainframes, and computers are systems with a physical location. Therefore cyberspace is a physical, social-technological

environment—a separate domain but one that interacts and blends with other domains and layers.[46]

This definition has the advantage of including both the communicative and the physical aspects of cyberspace, each of which matter for competition between states. The "cloud," as computer specialists say, is simply someone else's computer.

Other scholars have identified cyberspace as a new "commons," framing it alongside space, the air, and the sea.[47] A commons, as defined by Barry Posen, is an "area that belongs to no one state and provides access to much of the globe."[48] The term seems particularly appropriate for cyberspace, as many nations and other actors use the cyber commons for commercial, political, and military purposes, none of which necessarily exclude the others. Cyberspace is for the most part consensually communitarian; it is possible for the entire world of Internet users to access it simultaneously.

As with the other commons, cyberspace has been subject to attempts of a legal regulation. The development of a legal framework for managing the Internet, and for managing relations between individual and corporate actors in cyberspace, only began in the 1990s.[49] Realization of the perils and opportunities of cyberspace for the protection of intellectual property emerged in the late 1990s, as courts and legislators began to accommodate themselves to the unfamiliar realm, its limitations, and its capabilities.[50] Nevertheless, by the 2000s it had become clear that certain strategies of appropriation ran counter to national law (often in both attacker and target countries) and international law.

Many studies have explored the founding of the Internet and the parallel growth of legitimate and illegitimate Internet traffic.[51] By most accounts, the militarization of cyberspace happened prior to mass public awareness of the Internet, largely as a result of military interest in, and influence upon, the developing network.[52] Over the past decade, however, scholars and policy makers have increasingly concentrated on the prospects for, and implications of, "cyberconflict." Valeriano and Maness define cyber conflict as "the use of computational technologies for malevolent and destructive purposes in order to impact, change, or modify diplomatic and military interactions between states,"[53] and cyber espionage as "the use of dangerous and offensive intelligence measures to steal, corrupt, or erase information in the cybersphere of interactions."[54]

According to Valeriano and Maness, almost half of all cyber incidents involve theft, in which one state attempts to appropriate some kind of information from another.[55] Cyberespionage offers a low-key, nonconfrontational way for states to compete with one another over existing points of dispute.[56] States accomplish three goals by engaging in cyber espionage: they demonstrate capabilities, balance against rivals, and attempt to shift economic costs to competitors.[57] For the purposes of this chapter, we are most interested in how one state attempts to appropriate IP under the legal protection of another state—this is to say, when a state or the agents of a state use access to digital space in order to steal legally defined property. This process, which we will define as "cyber espionage" for the purposes of this study, runs counter to how domestic and international legal protection of intellectual property has developed.

The Basics of Cyber Espionage

Cyber espionage employs many of the same techniques as cyber crime, which unsurprisingly makes it a moving target for state intelligence and law enforcement organizations.[58] Unlike cybercriminals, however, cyber spies undertake cyber espionage in order to acquire information covertly, and as a result do not generally engage in such cyber-criminal tactics as vandalizing websites or initiating dedicated denial of service (DDOS) attacks, except insofar as these efforts support their main objectives. As a result, scholars and specialists in cyber espionage have identified several common techniques for entering target networks, and for illicitly appropriating information. Cyber espionage can also involve more sophisticated operations designed to cause extensive damage rather than simply steal secrets. These sabotage efforts—capable of doing severe damage to computing networks, to intra-organizational communications resources, and in some cases to physical infrastructure—skirt the line between traditional espionage and outright military conflict, as they can cause significant economic damage to the target.[59] Sophisticated attackers with considerable resources, such as state military bureaucracy, normally have access to a wide array of computers from which to launch attacks. To make attribution difficult (desirable to avoid both some simple defensive techniques and retaliation), attackers will often use techniques designed to disguise their country of origin.[60]

Cyber security specialists have struggled to stress the danger these techniques pose to private and government actors.[61] Unfortunately, compliance with best practices of cyber security can cost individuals and organizations more time and money than they are willing to spend. Consequently, confirming best practices across several linked organizations—for example, the Department of Defense, a major defense conglomerate, a range of subcontractors, and a series of law firms defending the interests of each actor—can prove exceedingly difficult even under optimal circumstances.

Given the difficulties associated with establishing any sort of defense against cyber attacks, it is not surprising that the world's militaries have had trouble establishing their own cyber war infrastructures. The integration of new technologies and concepts into existing military organizations often proves time-consuming and disruptive; the waves of disruption caused by the introduction of military aircraft more than a hundred years ago continue to ripple today.[62] The United States, among other countries, developed an overarching cyber security strategy slowly, and continues to struggle with organizational dynamics down to the present day.

One of the least tractable problems with creating an effective cyber security infrastructure is the difficulty of locating full responsibility within any single organization.[63] Every modern bureaucratic organization, whether military, civilian government, or corporate, requires defense against cyber attack, meaning that every organization needs to develop the capabilities of managing its own protection. This inevitably leads to different practices, varying tolerance for risks, and poor communication between organizations. It also means that the public can hold no single organization fully responsible for cyber security failures, not even for ones as large as the OPM hack. However, government can simplify the problem by centralizing some responsibilities, especially on the offensive side, in specific commands.

And yet the Obama administration began to develop a response to cyber-espionage in 2013.[64] In February 2016, President Obama followed this up with the proposal of a national action plan for cyber security.[65] The proposal included

- establishment of a Commission on Enhancing National Cybersecurity, which would strengthen cyber security in the public and the private sector, as well as promote cyber security technologies and best practices;
- creation of a $3.1 billion information technology modernization fund;
- development of a cyber security awareness program intended to improve standards and practices in everyday Americans; and
- a 35 percent increase in federal cyber security spending.[66]

These measures suggest concern over the vulnerability of the public-private nexus in the US technology economy. Steps to protect the private and the public sectors (by increasing the difficulty of stealing passwords, for example) go hand in hand, reinforcing the idea that vulnerabilities in one part of the network weaken the entire system.

In 2018 the Trump administration released its own comprehensive cyber security strategy. The strategy has four pillars, including

- defense of the homeland by protecting networks, systems, functions, and data;
- promotion of American prosperity by nurturing a secure, thriving digital economy and fostering strong domestic innovation;
- preservation of peace and security by strengthening the ability of the United States—in concert with allies and partners—to deter and, if necessary, punish those who use cyber tools for malicious purposes; and
- expansion of American influence abroad to extend the key tenets of an open, interoperable, reliable, and secure Internet.[67]

The Trump administration's cyber strategy specifically challenges some of the assumptions of the previous strategy, arguing that faith in the liberating power of information and free media was misplaced, and that cyberspace represented a critical area of contestation with adversaries such as Russia and China.[68]

CYBER STRATEGY AND THE PUBLIC-PRIVATE PARTNERSHIP

Former Secretary of Defense Ash Carter elaborated on the Defense Department's view of cyber security in an April 2015 speech at Stanford University (alluded to in chapter 3).[69] The location was not accidental, as the speech concentrated on the need for cooperation between the Pentagon and Silicon Valley. Carter argued that civilian innovation goes hand in hand with government action on cyber security. This relationship, which remained essentially the same under the Trump administration, has four aspects:

1. The increasing role that civilian investment plays in military technological innovation demands closer ties between the Department of Defense and the centers of civilian innovation.
2. Government investment and support have facilitated the development of many technologies central to digital innovation over the past several decades.
3. Private firms and the government face different facets of the same cyber security problem, as espionage threats target both private and public sector entities.
4. Private cyber security and publicly provided cyber security overlap; the defense of each depends on the security of the other, as both can come

under attack, and vulnerabilities in one sector can lead to vulnerabilities in the other.

Yet some argue that the public-private partnership that interests Carter and the rest of DoD is particularly unlikely to develop in the information technology sector.[70] Despite the critical role that government investment played in the foundation of the computing industry and the Internet, technology firms and their workers tend not to share the values of the military-industrial complex or have much interest in securing government contracts. The Pentagon, operating under government employment restrictions, cannot compete with Silicon Valley salaries. Moreover, the revelations of Edward Snowden exacerbated a long-term political distrust between the government on one side, and left- and libertarian-leaning tech workers on the other.[71]

The distrust between the Pentagon and Silicon Valley mirrors the problems that the Department of Defense has faced in broadening its procurement base to civilian-oriented firms. In the case of cyber defense, however, the problem is even more serious; DoD needs the active cooperation of technology and software firms in order to carry out its cyber security strategy. If private firms are vulnerable to espionage, then the Department of Defense cannot defend its system of procurement, or its basic military secrets.

THE US–CHINA CYBER ESPIONAGE RELATIONSHIP

While states and nonstate actors have engaged in a variety of activities that might fall under the umbrella of "cyber conflict" over the past decade, competition between China and the United States bears the most attention.[72] The United States and China have developed an enormous trade relationship since the 1980s, with investment spreading across nearly every sector of both economies. US and Chinese firms have engaged in dual production, with technology transfer, in an array of different fields.[73] Even big industrial producers such as Boeing have begun to work with Chinese manufacturers and subcontractors, transferring technology to their partners as necessary.[74]

The US government has periodically expressed concern about the amount of technology transfer to China, especially of potential dual-use technologies; but prior to 2017 actual interventions in private deals remained rare.[75] The problem goes to the very core of US foreign policy towards China. American firms seek to develop relationships with their Chinese counterparts in order to enter new markets, take advantage of relatively cheap Chinese labor and resources, and increase profits at home and abroad. Built on this foundation, the economic relationship between the US and China has become the chief dynamic in the Asia-Pacific region and in the world as a whole. Moreover, the financial aspects of this relationship have become central to the functioning of the world economy.[76] The Trump administration had a different perspective on this relationship.

Attribution for cyber attacks has rarely been a major problem in the US-China relationship, in part because the reputed crudeness of hacking efforts.[77] According to US cyber-experts, Chinese hacking activities began in the 1980s, with a substantial upsurge around the turn of the century. Experts began to suspect in the mid-1990s that Chinese hackers were secretly attempting, through a variety of cyber attacks, to appropriate the technology of American firms working in critical strategic fields. Many early efforts focused specifically on economic espionage, without much attention to either military or dual-use technologies.[78]

By the early 2000s, details began to emerge of persistent Chinese efforts to attack US computer networks, both in government and industry. We have no reliable account of the number of intrusions made by Chinese actors, as both government and private sources hesitate to release full details. However, John Lindsay and Tai Ming Cheung have assembled a list of thirty-seven intrusions into US networks between 2003 and 2013, including several large-scale attacks against defense-oriented networks.[79] Notable attacks included a 2006 intrusion into the US Department of Commerce, targeting export license data. In 2007, the "Ghost Net" attack targeted a wide variety of firms and public entities across 103 countries.[80] Another 2007 intrusion hacked BAE, Lockheed-Martin, and Northrop Grumman, acquiring nonclassified data on the F-35 Joint Strike Fighter. The 2009 "Elderwood" intrusion involved a sophisticated attack against defense and manufacturing firms in the United States. In the same year, a Chinese APT, or hacking group, called Hidden Lynx attacked hundreds of firms in the defense and financial industries. US industries are not the only targets; the "Luckycat" attack of 2011 concentrated on targets within the Indian defense industry, as well as Tibetan activist networks. "Nitro," another 2011 effort, attacked forty-eight chemical and defense firms. Another APT, Beebus Mutter, concentrated on drone technology in the United States and South Asia in 2011.[81]

In 2013, a report from the cyber security firm Mandiant argued that the People's Liberation Army has played a central role in China's cyber-espionage efforts, with what amounts to the official sanction of Chinese government authorities.[82] According to Mandiant, a unit associated with the PLA launched attacks against 141 global firms, many operating in the defense sector.[83] Reports indicate that these attacks have sought draft patent information, organizational strategy and hierarchy, and trade secrets.[84] According to FBI Director James Comey, there are "two kinds of big companies in the United States . . . those who've been hacked by the Chinese and those who don't know they've been hacked by the Chinese."[85]

The participation of an active-duty PLA unit in efforts to steal US defense-sector-related IP indicates that state behavior in the field of espionage is adapting to new technological and legal realities. The strategic relevance of cyber crime becomes tied to the rise of IP protection as a critical national concern. To the extent that the Chinese government facilitates the appropriation of important US intellectual property, especially in the defense industry, it threatens the national security of the United States.

Defense Material

Chapter 3 detailed the deeply complex set of relationships between contractors, subcontractors, law firms, and government that characterizes the contemporary US defense industry. The multiplication of the number of independent players in the US national innovation system (a trend mirrored in other countries) means that cyber attacks can acquire critical defense information from a wide range of sources, both public and private. As chapter 3 discussed, virtually every patent owned by traditional national security providers in the United States involves de facto collaboration between the US government and a private actor.

The data revealed by Edward Snowden and other sources indicates that the United States believes that China has appropriated a considerable amount of technology associated with numerous defense systems including the F-35 Joint Strike Fighter and the General Atomics MQ-1 Predator drone.[86] Of these, the F-35 represents the most critical vulnerability. The Department of Defense expects the F-35, a product of Lockheed Martin, to fill out the fighter-bomber fleets of not only the US Air Force, Marine Corps, and Navy, but also the fleets of nearly a dozen allied states.[87] Classified presentation slides published by *Der Spiegel* indicate the loss of detailed information regarding radar design and engine schematics, as well as terabytes of engineering and testing data. These slides also indicated a US government belief that it had suffered[88]

- 30,000 incidents,
- 1,600 computers penetrated,
- 60,000 user accounts compromised,
- 33,000 US Air Force field officer records,
- 30,000 US Navy passwords, and
- information on

o air refueling schedules,

o the US Transportation Command (TRANSCOM) single mobility system,

o US Navy missile and navigation systems,

o US Navy nuclear submarine and anti-air missile designs,

o International Traffic in Arms Restrictions (ITAR) data, and

o data on the B-2, the F-22, the F-35, and other aircraft.

With respect to the F-35 in particular, China could use technical information appropriated from the project in several ways. First, it could apply technical know-how to efforts to detect and defeat the F-35. This would involve improving the capabilities of Chinese detection and weapons systems in ways that could ensure detection, and a successful kill following detection. Potentially, China could share this information with other interested states, just as the United States, Israel, and others shared information about Soviet MiGs with one another during the Cold War.[89] This would fall into a traditional understanding of military espionage, and would not represent a significant violation of the intellectual property rights of US firms. Second, China could use the appropriated technical information to improve its own jet fighters, potentially competing with US aircraft. Some indications suggest that China is moving in precisely this direction. The J-31 fighter prototype, produced by the Chinese military aviation firm Shenyang, reportedly has many features that Shenyang has directly copied from the F-35.[90] The two aircraft are not identical, as the J-31 has two engines to the F-35's one, and the J-31 lacks the architecture for VSTOL flight that is central to the F-35. Nevertheless, some similarities suggest that Shenyang had access to proprietary information about the F-35 when designing the J-31.

Adding to the complication, recent reports have indicated that China plans to export the J-31, and indeed that Shenyang may build the aircraft primarily for export, rather than for the domestic market.[91] This would put the J-31 into direct competition with the F-35 as the only fifth-generation stealth fighters currently available. Potentially, this could open Shenyang and the Chinese government up to legal action under several instruments of international intellectual property law. While it is unlikely that the United States could enforce a settlement inside China, a ruling could potentially affect Shenyang's assets abroad, making them subject to lawsuits or other legal sanction.

More broadly, the evidence suggests that Chinese cyber espionage has targeted technical subsystems that can be incorporated into broader weapons projects.[92] These include dual-use technologies that often fall under US export control prohibitions. The targeting of dual-use technologies further expands the number of firms that could come under attack from Chinese cyber-espionage assets.[93] Again, the nature of the modern military-industrial complex, reliant on the defense firm as aggregator and on a broad base of civilian-oriented technology firms, exacerbates vulnerability to espionage.

However, the extent to which such data has found its way into Chinese military technology remains uncertain. All the problems associated with traditional industrial espionage, and particularly with reverse engineering, apply to espionage with digital tools. Successful espionage requires several sequential steps.[94] The PLA must communicate its technological needs to China's defense industry. The defense industrial units must assess their areas of weakness, and request the theft of information by the PLA and by China's intelligence agencies. These agencies must identify the appropriate targets of espionage and gain entry to them in a timely fashion. It must then pass this information back to the appropriate consumers within the defense industry, who must absorb the stolen data into their own systems of research and development. This process can break down at any number of points, resulting in a failure to acquire and deliver relevant information to the appropriate actors in a timely fashion.

By contrast, the kind of cyber espionage undertaken by US intelligence agencies has different implications for intellectual property protection. As discussed in the introduction to this chapter, the US intelligence community (IC) has historically supplied US firms with a variety of intelligence designed to improve their market position and negotiation strategies.[95] The IC has also fed intelligence about foreign military equipment into the private sector of the defense industrial base, although it is less clear that this intelligence gathering has directly affected the ability of US firms to compete with foreign products.[96]

The US Response

Much about the US response to these cyber attacks remains unknown. The revelations of Edward Snowden made clear that the United States does undertake sophisticated cyber espionage efforts against China and Hong Kong.[97] According to Snowden, the PRISM program, under the aegis of the National Security Agency, collects information from a group of nine US Internet firms. This information includes chats, emails, and other direct forms of communication between companies, individuals, and the government, and it gives the United States legal access to much of the information that Chinese hackers need to steal, as well as insight into hacking activities.[98] The US government may also undertake other even less visible measures to deter or retaliate against Chinese hacking.

The United States has also responded through a combination of diplomatic and legal tools. The Federal Bureau of Investigation, among other organizations, has employed sophisticated forensic computing methods to trace incidents of espionage to Chinese hackers.[99] In May 2014, the US Department of Justice indicted five officers of the People's Liberation Army on charges of cyber theft.[100] The US indictment established an important distinction between US espionage policy and Chinese policy. The key difference, according to US policymakers, is that the PLA hackers stole information from private US firms and turned it over to Chinese state-owned firms. Recall that US espionage purportedly concentrates on governments and state-owned firms. While the information gained from such espionage may benefit private American firms, it does not involve the straightforward transfer of foreign information to privately-owned American companies.

In the belief that private Chinese companies often benefit directly from both state-supported and privately conducted industrial espionage against US firms, the Obama administration issued an executive order in March 2015 that individual Chinese companies would be subject to financial and legal reprisal.[101] Some experts worried that this would deter US companies from working with Chinese counterparts, for fear of inadvertently transferring key intellectual property. Other experts suggested that China might retaliate against the United States by selectively enforcing regulations on cyber security to American firms operating in China.[102]

Many critics suggested that these indictments would have no meaningful effect on Chinese cyber activities, largely because the United States could not deliver or enforce the indictments on Chinese soil.[103] However, changes in China's approach to cyber security suggested that the indictments may have had a more significant effect than initially expected.[104] Following the indictments and a broader flurry of US complaints, President Xi reportedly dismantled some elements of the PLA's suite of cyber-warfare capabilities. This included a crackdown against "moonlighters" who used PLA spying as a cover for appropriating US intellectual property and selling it to Chinese firms. According to the *Washington Post*, since 2014 Chinese hacking largely shifted from the PLA to the Ministry of State Security, which undertakes more traditional spying efforts.[105] Still, in December 2015, China arrested several individuals suspected of participating in the hack of the US Office of Personnel Management.[106]

In 2018, the Trump administration took a harder line on Chinese IP appropriation. Spurred by a DiUX report on China's espionage activities, the United States threatened aggressive sanctions against China, and against specific Chinese firms. The DIUx report detailed cyber espionage, collection of open source information, development of human capital in US universities, and forced technology transfer by US firms.[107] The report recommended domestic reforms in the United States, more aggressive steps toward countering Chinese activities, and specific punishments for China. At the time of this writing, the fallout from the report remains uncertain. And in 2019 the Trump administration took its most aggressive step yet, sanctioning the technology firm Huawei. Concerned about the foundational position that Huawei might achieve in the development of 5G networks, the Trump administration banned US companies from working with the Chinese firm. This essentially prevented Huawei from accessing the patents it uses for much of its equipment, even with respect to components produced outside the United States. Because of the intertwined nature of technology supply chains and the myriad of different IP claims within those chains, the Trump administration's actions have the potential to severely disrupt Huawei's operations.[108]

Political Pressure and Cyber Restraint

Why did China, at least briefly, prove flexible on cyber-conflict management? The answer may lie in some basic interpretations of the history of cyber warfare. Many analysts have predicted that the opening of cyberspace would lead to national conflict, and to government conflict against subnational groups. However, Valeriano and Maness argue that the chief characteristic of conflict in the cyber age has been restraint.[109] While the development of the cyber commons opens up wide avenues in which states can attack one another, most governments thus far have not pressed their advantages. In part because of uncertainty about their own vulnerability, states restrain themselves from escalating. Some states may also worry about the principal-agent problem: the degree to which they can exert full control over the cyber capabilities they develop and the individuals who operate them. And while nonstate proxies may offer some states certain advantages, these states may also be concerned about the use of such groups because of the long-term threat that enabling them could pose.[110]

Viewed in this context, the Chinese response to American pressure fits into a general pattern of international behavior. If the theory of cyber restraint proves correct, then legal and diplomatic efforts at preventing the improper appropriation of intellectual property may yet bear fruit. The nature of restraint lies in the development of norms of appropriate behavior. International and domestic law help to inform and produce these norms. Conceivably, careful pursuit of international IP protection could put bounds on the acceptability of some forms of cyber espionage. Thus, intellectual property law could help determine how states think about the appropriateness of certain kinds of cyber espionage under certain circumstances. If US officials have their way, the bounds would involve respect for the property rights of private firms, and a renunciation of efforts to directly copy and export foreign systems.[111] Of course, the US decision to destroy Huawei may undermine efforts to develop sustainable norms with China.

If cyber restraint comes at least in part from queasiness about principal-agent problems, then the structure of cyber weaponization may matter for how states behave. At the same time that China has embarked on an apparent reevaluation of its cyber operations, President Xi has spearheaded a massive reorganization of China's military establishment.[112] This could bring China's corps of cyber soldiers, which has apparently developed in ad hoc fashion, under tighter central government control. However, it could also create a constituency within the military, and consequently within government, for more aggressive use of cyber espionage.[113]

Of course, the United States faces some of the same issues. Currently, responsibilities for cyber warfare exist across a range of military and intelligence agencies. In the Department of Defense, the US Cyber Command coordinates efforts across the services.[114] In the intelligence community, the National Security Administration has taken the lead on cyber security, though other organs of the IC also maintain capabilities commensurate with their missions. Some have argued that the United States should centralize its military capabilities, at least, in a service dedicated to cyber warfare, somewhat akin the development of the US Air Force in 1947.[115] However, the prospects for such reform seem dim.

The 2015 summit between US President Obama and President Xi of China seems to have borne some fruit with respect to cyber espionage. In the summit, Obama reportedly expressed the gravity of US concerns over China's hacking activities against both civilian and military targets. The summit resulted in a joint statement pledging that "neither country's government will conduct or knowingly support cyber-enabled theft of intellectual property, including trade secrets or other confidential business information, with the intent of providing competitive advantages to companies or commercial sectors."[116] Xi followed up these statements with similar commitments to Germany and Great Britain, and participated in a multilateral statement to the same effect at the G-20 conference that year.[117] Of course, the notable shift in US rhetoric following the election of President Trump very possibly undercut even the limited progress that the summit may have achieved.

Given the danger posed by unconstrained cyber conflict, the technology policy specialist Fergus Hanson has suggested developing norms that allow differentiation between "military attacks during war, state-backed economic espionage, political espionage, and military-style attacks during peacetime."[118] Hanson argues that states need to establish a clear set of standards for how to respond to each of these attacks, as they all carry complications. The latter two kinds of attack invite military escalation, as they threaten a state's ability to manage itself and protect its population. The first kind seems simple, but the level of response may depend on the severity of attack and have synergistic effects on other battlefronts. But if US pressure against China has really had an effect on the latter's conduct of state-based economic espionage, this may suggest genuine progress in creating norms of appropriate cyber conflict. Again, however, the Trump administration has pursued a different strategy.

Many questions remain, in no small part because of uncertainty about China's long-term intentions.[119] The appeal of cyber restraint may decline over time, as states become more sure-footed with regard to their capabilities for cyber warfare. Moreover, China may come to resent the asymmetrical impact of the norms, which protect US, European, and Japanese technology more than Chinese.[120] If so, the immense productivity that the cyber commons has enabled may yield to a set of entrenched, protected fiefdoms.

Conclusions

Many of the implications of the cyber revolution remain difficult to ferret out.[121] The expansion of IP law has the potential to create contradictory effects on international espionage. On the one hand, the accumulation of data, ease of cyber access, and proliferation of actors makes it easier for hackers and spies to acquire intellectual property. On the other hand, the growing acceptance of international IP law may offer victims a new set of instruments for fighting espionage.

The People's Republic of China has pursued industrial espionage as a defense innovation strategy since its founding. Only the methods have changed. Early in the Cold War, the Chinese attempted to copy complete systems, or at least learn as much as they could from subsystems. Later, after the Sino-Soviet split, it produced unlicensed versions of Soviet equipment, from tanks to submarines, aircraft, and assault rifles. The Chinese military industrial complex steadily worked incremental innovations into those platforms, but the lack of investment, along with assorted political difficulties associated with the Cultural Revolution, meant that China fell further and further behind both the United States and the Soviet Union.

After the Sino-American rapprochement China gained access to some Western technology, which helped facilitate its incremental strategy. Only in the 1990s, however, when China gained access to the latest Russian technology, did the Chinese national innovation system kick into high gear. At the same time, the creation of a system of intellectual property protection internal to the DIB helped improve the innovative climate (as discussed in chapter 4). China appropriated technology it had purchased from Russia, reverse engineering and producing several major platforms, though the lack of access to trade secrets limited the effectiveness of this espionage.

In the last decade China has gained access, through digital means, to a wide array of Western intellectual property, including civilian, military, and dual-use innovations. At the same time, the Chinese DIB has diversified beyond its state-owned core to embrace a family of civilian technology firms. It now appears that these firms are working together with the People's Liberation Army to facilitate the acquisition of US military and dual-use technology for military purposes.

The role of intellectual property IP law has been to shift the terms under which China engages in industrial espionage, and potentially to increase its efficacy. While China may have the most formalized, virulent cyber espionage program, the scope of cyber theft of IP is global. Part of the problem was generated, ironically, by the expansion of IP law itself. As managing IP protection in the modern economy has become more complex, the number of actors needing access to information has increased. Cyber espionage can target any of these actors, seeking weak links to appropriate critical properties. The digitization of knowledge makes the business of IP legal protection easier while at the same time making the information more vulnerable to cyber theft.

At the same time, domestic and international IP law offers answers. To the extent that the international IP regime can drag such states as Russia, China, India, and Iran into its embrace, it can create the foundation of domestic legal systems that can constrain bad behavior. International IP law can also offer weapons with which to fight certain kinds of espionage, particularly on the part of actors with some kind of international exposure. Finally, states are actively seeking norms of restraint in cyber espionage, and international IP law can help provide those norms.

7: CONCLUSION

Vignette: Rise of the Drones

The MQ-1 Predator drone has become the unlikely face of American air-power. An unmanned aerial vehicle (UAV), the Predator became the center-piece of the long-running campaign against Al-Qaeda, the Taliban, ISIS, and various other groups affiliated with jihadist ideology. Predator drones, along with the follow-along Reaper design, conduct reconnaissance and carry out air strikes around the world.[1]

The history of unmanned aerial vehicles (commonly, although incorrectly, known as "drones") stretches back to the aftermath of World War I.[2] The development of purpose-built drones for reconnaissance and attack missions by the United States Air Force began in the 1950s, as the service worked through the implications of unmanned ballistic and cruise missiles.[3] The first deployable US utility drones emerged out of cooperation between the Air Force and the National Reconnaissance Office in the early 1960s.[4] The NRO provided the institutional vehicle for collaboration between the Air Force and the Central Intelligence Agency, as the latter was interested in the intelligence-gathering capabilities of drones.[5]

By the late 1970s, developments in Soviet ballistic missile technology such as rail-mobile ICBMs created a demand for reconnaissance assets with persistent loiter capability (aircraft that could remain on station longer than pilots could fly them). Neither the SR-71 (the most advanced high-speed spy aircraft) nor the satellites of the time could provide this capability.[6] Research into a drone that could fill this gap did not result in a useable vehicle, but did keep money flowing into the program. By the 1980s, the Air Force began losing interest in drones. Stealth aircraft such as the F-117 and the B-2 could strike Soviet targets during a general war, and did not suffer from data link vulnerability.[7] The need for operators to stay in contact with drones made them unreliable in high-intensity warfare and other critical situations. Air Force planners worried that the Soviets could exploit a UAV's need for a sophisticated data link during critical strike missions. Cruise missiles, which did not rely on such links, also appeared more attractive.[8] In any case, the prospect of having sufficient bandwidth to communicate complex data (video, photographs, and flight instructions) seemed distant during the 1970s and 1980s. By the early 1990s, US Air Force research into UAVs had virtually ceased.

Then, in 1993, the Defense Airborne Reconnaissance Office (DARO) took over control of UAV research and development, and helped shepherd the General Atomics RQ-1 Predator drone into existence. Created to answer the need for long-term loiter reconnaissance during the Bosnia conflict, the General Atomics Predator used civilian satellites for data uplink, and took advantage of the Global Positioning System (GPS).[9] The Predator served over Bosnia, in the Kosovo War, and in Operation Southern Watch (surveillance over Iraq). It would serve as a forerunner for such craft as the RQ-4 Global Hawk and the MQ-9 Reaper. In 2001 a program to equip MQ-1 Predators (the designation shifted from "R" to "M" with the armament) with Hellfire missiles came to fruition, and the UAV close-air-support mission began.[10] The first "drone strikes" were launched in October 2001.[11]

The effectiveness of the Predator depends on several related but distinct technological trends. In part because of the unpredictable development of these technologies, the Predator emerged outside the normal US defense procurement process. The airframe, which is limited in speed, size, and maneuverability relative to most military aircraft, is in some ways the least important part of the system. Far more important is the data link, which allows UAV operators to remain in consistent contact with the vehicle while receiving and processing huge amounts of information.[12] Prior to the 1990s, communication and computer processing technology had not matured sufficiently to make the UAV an effective collector of real-time intelligence. Operators ran the risk of losing control of the aircraft, and therefore of a crash, possibly in hostile territory. At the very least, drone operators could not feed information directly from the aircraft to customers (shooters, policy makers) down the line.[13] But, as computer and communications technology developed—with a jump start from government, but mainly through civilian investment—the prospect of controlling a Predator became a lethal reality.

In short, the Predator and its kin exist because of the confluence of civilian and military investment, with dual-use technologies at its core. These technologies, often developed by small, civilian-oriented firms reliant on intellectual property protection, made the Predator an effective military platform. While the US government supported and invested in many of these technologies, the firms that produced them generally focused their attention on the civilian market. Computer processors, software, and wireless communication technology enabled the rather milquetoast Predator airframe to become the face of American airpower during the War on Terror.

Also reflective of changing realities in the defense industry, the Predator emerged as the result of a partnership between several firms, not all of which had historically focused on the defense market.[14] Stemming in part from the need for inclusion of dual-use technologies, and in part from changes in the organization and structure of the defense industrial complex, the Predator became emblematic of a trend in defense innovation that saw small firms develop partnerships with large, traditional defense conglomerates. These partnerships require careful negotiation of intellectual property rights, as each firm accepts some risk from working with the other.

This system places stress on the procedures the US government uses to secure its access to military equipment and protect intellectual property. Although civilian-oriented and partnership-seeking firms want control over their own IP, they require active legal protection from the US government in order to achieve this. The US system of IP protection, in effect, guarantees the terms of defense partnerships, as well as the rights of civilian-oriented firms to their inventions. The prospects of such partnerships, and the interest of civilian firms in participating in the defense market, depend on the nature and extent of such protection. In this sense, the MQ-1 Predator exists because of a specific configuration of intellectual property.

The apparent effectiveness of the MQ-1 Predator has spawned clones around the world.[15] Russia, China, and several other countries have developed drones similar to it. Indeed, China appears to have what amounts to a clone of the Predator.[16] In addition, its CH-4 armed UAV strongly resembles the General Atomics MQ-9 Reaper, a larger, more advanced version of the Predator operated by the US Air Force.[17] Bill Gertz, working from a video of the CH-4 posted online, writes: "Both aircraft are about the same size and wing-span and both sport identical V-tails, landing gear, imaging pods and propeller-driven rear engines."[18] To be sure, no one can yet prove that China acquired information about US drones through illicit means from General Atomics, the Department of the Defense, or the myriad of contractors, subcontractors, and law firms associated with the development and sale of US weapons. The Reaper depends for its success on a set of technological advances in computing and communications that China already has access to through the civilian market. As chapter 6 discussed, much of what looks like industrial espionage actually involves parallel development, sometimes supported by legitimate, open-source acquisition of technological innovation. Put differently, even without secret data about the Reaper and other US drones, China could likely construct an aircraft of similar capabilities. That said, some customers for Chinese drones have expressed discontent with

Chinese quality control. Jordan, for example, attempted in 2019 to resell its CH-4 drones after only two years of use.[19]

The potential theft of proprietary technology may matter on the export market, however. Since 2013, China has become a major exporter of UAVs.[20] If the US Department of Justice could ever conclusively demonstrate that Chinese hackers stole US technology, the firms that produce the drones could conceivably come under sanction, especially if they operate in countries friendly to US legal intervention. And that could make buyers uncomfortable enough to hesitate before pulling the trigger on a big arms deal, especially if they represent states that have come under the umbrella of the emerging international intellectual property regime.

Introduction

Isolating the impact of intellectual property protection on the innovation, production, and diffusion of military technology is a research program necessarily in its early stages. While IP issues represent a fast-growing sector of economic statecraft, research into how economic statecraft interacts with defense statecraft remains sparse. However, this book has demonstrated that IP concerns have an impact at every stage of the innovation-diffusion-adoption process. First, the structure of IP law can affect the ability of states to innovate, structuring the defense industrial base to meet specific needs and develop certain types of military technology. Second, concern over respect for IP can alter how a state exports particular kinds of arms. Finally, in combination with changes in information technology and the structure of the modern military-industrial complex, the relevance of IP to modern military technology has opened up new avenues of international espionage.

As Susan Sell has suggested, international IP protection is, in and of itself, a power play on the part of major economic actors.[21] The construction and maintenance of the rule systems owes itself to the entrepreneurial behavior of private business, working not only through the US government but also through international institutions. As such, power relations are embedded within the rules of the IP system, and within our entire way of talking about intellectual property. This is one reason why the IP provisions of the proposed Trans-Pacific Partnership, for example, proved so controversial.

But adherence to international institutional frameworks is not entirely voluntary. The demands of international organizations (and, in bilateral terms, of the European Union and the United States) require the Chinese government, for example, to develop a position on intellectual property, a set of policies designed to support that position, and the bureaucracy necessary to execute those policies.[22] While this bureaucracy may lack power initially, over time the state acquires what amount to habits of compliance, where it becomes more problematic to step outside the expectations of the international regime than to stay within them.[23]

Summary of Findings

This study has synthesized several large bodies of work on the nature of intellectual property protection in the defense sphere. Intellectual property law has affected the pursuit of military advantage for virtually the entire history of the modern state. Rulers in the early modern period used patent protection to support the development of military technologies that would allow them to maximize national security. These efforts helped lay the foundation for modern intellectual property law. Indeed, at the opening of the twentieth century, IP law played a central role in creating the boundaries and structure of the modern military-industrial complex, helping business and the government form stable relationships in defense procurement and military technology development.

With respect to the United States in particular, intellectual property law holds a central position in the modern American defense industrial base. This system consists of a vast array of defense-oriented private firms, government agencies, military organizations and offices, independent labs, universities, contractors, subcontractors, law firms, and civilian-oriented producers of dual-use technology. It depends on intellectual property law to guarantee ownership, attribution, and secure transfer of technical innovations. The Invention Secrecy Act and the state secrets privilege continue to have a major impact on how these actors function, and on what they can expect when they try to negotiate lasting commercial agreements.

As intellectual property law has become critical to the functioning of the modern economy, it has spread across the international system. Three centuries ago, states—especially those trying to catch up with leading industrial nations—bitterly resisted the concept of international IP protection, formulating policy around the idea of appropriating foreign technology. Today, the United States and a group of advanced Western countries have managed to develop and spread a system of legal IP protection that involves an ever larger part of the global economic system. This has created obligations for states (including those trying to catch up in military technology) that they would not have suffered a century ago.

Intellectual property regimes have also helped structure national innovation systems outside the United States. Both the Soviet Union and China developed regimes of intellectual property protection based on idiosyncratic ideological foundations. These regimes, in conjunction with other institutional elements of their military-industrial complexes, helped to structure the nature of technological innovation in the defense sphere. As China and Russia have reformed and modernized the broader aspects of their regulatory economic strategy, they have also changed how they approach intellectual property in their MICs. This change has had significant implications for how they innovate, and for how they transfer military technology abroad.

The transfer of military equipment through arms sales has long been one of the central vectors through which military technology diffuses across the international system. Since the beginning of the Cold War, the United States has used a system of export controls to prevent the transfer of sensitive intellectual property to hostile foreign countries. This practice continues alongside a series of other practices that states have adopted to manage the sale of their IP to customers. Licensing, coproduction, and technology transfer agreements now characterize virtually every major licit global arms deal. The transnational integration of production and innovation have complicated the picture, requiring further interventions on the part of intellectual property law. The increasing importance of dual-use technology to military innovation has complicated the situation even further.

The expansion of IP law, like many technological and legal developments, has the potential to have contradictory effects on international espionage. On the one hand, the accumulation of data, ease of cyber access, and proliferation of actors makes it easier for hackers and spies to acquire IP. On the other hand, the growing acceptance of international IP law may offer victims a new set of instruments for fighting this espionage.

Theoretical Implications

These findings suggest that political scientists have understated the role that the regulatory state plays in setting the stage for the development and diffusion of innovative military technology. The nature of the domestic regulatory state has observable effects on how the national innovation system produces and integrates novel military technologies. The impact of the regulatory state increases as the NIS becomes more complex, involving more actors who must protect their rights to their own contributions. The legal structure of intellectual property protection, manifested in both legislation and judicial procedure, affects the incentives of companies to contribute to the defense economy, and consequently imposes a limit on military access to innovative technologies.

Political science research into the diffusion of military technology may also have understated the impact of domestic and international legal structures. States—including those with a revisionist approach to international law, such as Russia—take seriously the potential for IP appropriation when making decisions about arms transfers. The United States regularly rejects potentially lucrative arms transfers to reliable allies because of concerns about the security of intellectual property. International legal structures are increasingly having an effect on this process, both by standardizing levels of IP protection and by creating mechanisms for exacting costs for IP violations. This increases the seller's confidence that its technology will remain secure, thus reducing impediments to the transfer of arms and arms technology.

There is little indication that the United States intended to "weaponize" intellectual property law when it helped sponsor the development of robust systems of IP protection in the 1990s.[24] This process was driven largely by corporations in the entertainment, agriculture, and pharmaceutical sectors, all of which sought tighter protection in order to guarantee their own profits against competitors in the developing world. This effort has been remarkably successful, as the TRIPS Agreement became binding on WTO members in 1989, and a flurry of bilateral treaties, all with extensive IP regulations, followed in the early twenty-first century.[25]

America's status as a first mover on intellectual property protection can obscure the fact that the system is at its foundation voluntary. The international community accepted the growth of the international IP regime because of pressure, but also because it saw advantage in promising to follow the rules that the United States had helped to set. As nations perceive IP law not simply as an obstacle to navigate but also as a potentially lethal weapon directed at the core interests of their major corporations, they may well become more reticent about accepting the preconditions that the United States wants to establish. The weapons of interdependence only work insofar as major players see advantage in allowing them to work. And that, in the end, means that efforts to exercise US power may result in the loss of that power. In short, turning IP protection into a lethal weapon the United States can use to kill companies it doesn't like runs the risk of causing a backlash and halting the momentum that produced strong protection in the first place.

Policy Implications: The Third Offset Revisited

To recap a theme visited in chapter 1 of this book, in the last decade the US military-industrial complex (led especially by the Pentagon) has concentrated on the idea of the "third offset," a program to develop innovative technology that would allow the United States to retain its military advantage over China, Russia, and others. This effort has included substantial outreach to private, civilian-oriented businesses, especially in the technology sector. The premise behind the third offset is the idea that the digitization of knowledge has made technological advantage over potential foes fleeting, and that therefore only a system of continuous innovation can maintain America's position.

As previous chapters have suggested, this question has potential implications for both domestic and international practice of intellectual property law. Beyond using international IP law to ensure US industrial access to the world's most advanced technologies and technology-producing systems, the United States can weaponize IP law to interfere with China's efforts of China to build competitive military technology. This weaponization involves taking the same steps that the US government has taken to protect nonmilitary forms of intellectual property: primarily, an expansion of bilateral and multilateral agreements to protect IP. These agreements help to establish an international standard for IP protection, even for states that do not formally abide by them or that do not expect to conform to their provisions.

Surely, the United States cannot convince Beijing or Moscow to prosecute violations of the intellectual property rights of US defense providers. Military technology remains a core state interest, and US competitors are unlikely to give up their most effective tools for keeping pace with US military technology. As discussed above, the United States has used these tools primarily to effect change in trade partners, such that prospective partners adopt US norms and restrictions on US intellectual property. For the most part, the United States has focused its efforts on "civilian" intellectual property, including copyrighted works of art, patented pharmaceuticals, and trade secrets associated with industrial production techniques.

The United States should explore the extent to which it can use these same tools to retaliate against the appropriation of intellectual property in the military sphere. Washington cannot convince Beijing to give up on industrial espionage, but it might be able to convince engineers working for Shenyang from attending international conferences in states where they might face legal retaliation for suspected violations of US patent and trade secret law. Perhaps more important, it could conceivably convince executives of Chinese civilian firms that supply Shenyang or Chengdu with the electronic components necessary to make their systems function that they need to be careful about who they work for, and when. The US government could also conceivably take action against firms in third-party countries that have agreed to the central tenets of US intellectual property protection.

Chinese companies, not to mention the Chinese government, need to work with foreign firms and governments that place a high priority on the protection of their IP. The need for the Chinese government and Chinese companies to navigate an increasingly complex international IP regime helps strengthen IP protection inside the country. From an administrative point of view, successfully managing the intellectual property aspects of a national innovation system is a tremendous challenge, especially for a culture without a strong intellectual property tradition, and a state built in express defiance of capitalist property norms.[26] Nevertheless, with the now-established connections Chinese firms have developed with their foreign-owned counterparts, the foundation for a robust IP system is potentially solid.[27] On both the civilian and the military sides, Chinese officials now appreciate that IP rules facilitate information transfer and sharing, and that a strong culture of sharing is necessary in order to promote widespread innovation.[28]

But a dual-track policy of management, in which the Chinese government attempts to preserve patent and trade secret protection for Chinese firms while exposing foreign firms to attack and expropriation, is difficult if not unworkable in the long term. The extremely complicated set of business and legal relationships that characterize the IP world will make a dual-track system impossible. This is not to say that all Chinese industrial espionage will end, but if China develops an internal system of guaranteed property rights, the norms and practices associated with these property rights will inevitably seep into its foreign technology acquisitions.

When considered under these terms, the pessimism about international technological diffusion that underlies much of the third offset requires some tempering. US firms in the civilian sphere stay ahead both because of their innovative characteristics and because they can protect those innovations through intellectual property law. We cannot expect that US defense firms will enjoy complete protection of their IP on the international stage, but we can expect that civilian firms focusing on dual-use technologies will enjoy considerable protection, and that even the big, traditional defense providers will be able to take advantage of some aspects of IP law. This expectation implies that the diffusion of military technology may not happen as easily or as seamlessly as some expect or fear. It also means, however, that intellectual property protection may become a core interest of the United States not only in economic terms but in the defense sphere.

The Trump administration's aggressive steps on intellectual property protection may herald this new perspective. The administration opened a section 301 investigation of China's IP practices in 2017, and used the results of this investigation to open a round of tariffs against China.[29] More recently, the administration took steps to cut the Chinese technology firm Huawei off from any US intellectual property, effectively isolating Huawei from international supply chains. This was undertaken in part because the administration believed that Huawei intended to manipulate its position at the core of the international technology economy to establish foundations for future IP espionage.

Reform

The United States needs to reform its approach to intellectual property in the defense sphere. The current IP practices of the Department of Defense make it more difficult, not less, to pursue the Pentagon's goal of diversifying the procurement base beyond the large traditional firms of the defense industrial base. Current IP practice puts small civilian-oriented firms in jeopardy with respect to their position in the civilian market, their legal vulnerability to the government, and their ability to conclude fruitful partnerships with defense-oriented firms.

First, the Department of Defense should reform its legal approach to the state secrets privilege, such that firms have reasonable legal recourse to infringement. Abuse of the state secrets privilege not only damages specific firms that have an interest in defense contracts, but also chills the enthusiasm of civilian firms for working in partnership with traditional defense providers. As demonstrated in *Lucent v. Crater*, under some circumstances suppression of evidence under the State Secrets Act may disregard inventors' constitutional rights. Where the government has directly or indirectly misappropriated a trade secret, the owner has a constitutional right to compensation for the taking. Suppression of the evidence needed to prove the owner's trade secret claim removes this constitutionally mandated remedy. To add salt to the constitutional wound, any direct takings claim would generally fall victim to the same problem.

Equally problematic is the deterring effect of the US government's destruction of inventors' IP claims that will fall disproportionately on smaller businesses. This directly conflicts with DoD's contention that smaller firms will need to produce a substantial share of the next generation of defense technology in order to maximize the nation's security. This result is also at odds with DoD's stated goals of encouraging small firms to consider it as a potential customer for technological innovation. Smaller firms, already facing substantial obstacles in breaking into the defense procurement market, lack the political influence and long-term institutional relationships to ensure that the military will properly license their technology. As a result, these firms must rely primarily on legal protection, rather than political or economic clout, to vindicate their intellectual property rights. If the application of the state secrets privilege strips these smaller inventors of this protection, they may be disinclined to pursue inventions sought after by the military. In essence, the nation's long-term defense strategy is being taken hostage by a short-term litigation strategy.

DoD should modify its IP acquisition practices to ensure that civilian firms feel securely in control of their intellectual property. This includes simplifying the procurement process such that these firms have greater confidence in the security of their property as well as a better understanding of how DoD approaches competitive bids. DoD should also take a more flexible stance on the acquisition of trade secrets and patent rights from civilian-oriented firms, recognizing that changes in the nature of defense procurement have rendered some traditional practices obsolete.

More broadly, the US government should stake out a position in international forums on best practices for handling defense IP. The defense firms, as discussed in chapters 3 and 4 of this book, have adapted slowly to the new intellectual property landscape, in part because it has not heretofore been necessary for them to adapt quickly. Consequently, they missed out on the ability to contribute to the formulation of the international IP regime, despite having strong interest in export, international supply chain management, and the minimization of piracy and IP theft. Yet the dominant US defense corporations remain in a uniquely advantageous position with respect to the formulation of international IP rules. First of all, the United States remains the world's largest arms exporter. US defense firms have subsidiaries and subcontractors all over the world, and they wield a great deal of influence on the security decisions of foreign governments. While the US government cannot prevent all instances of industrial espionage, it can jump-start the development of an international legal framework in which US defense producers can retaliate against the piracy of military IP. This framework can help replace the ad hoc arrangements that characterized US relations with its allies during the Cold War. It can also provide tools for fighting infringement by non-allies, largely by outlining measures for retaliation against individual, corporate, and state-based offenders.

The World Intellectual Property Organization has, in fact, already put into place programs designed to establish respect for IP law within the legal systems of developing countries. Although the Trans-Pacific Partnership has failed, and the future of the Transatlantic Trade and Investment Partnership remains uncertain, bilateral efforts will continue to shore up this multilateral structure. These efforts, though not driven by defense industry firms, may serve to make transnational integration of the global defense industry easier.

Avenues for Future Research

This study has broad implications for future research on how intellectual property law affects the innovation, diffusion, and production of military technology. Future studies can examine comparative innovation frameworks in greater detail. Industrial, IP, and multilateral arrangements differ widely across the OECD. European countries have defense priorities very different from those of the United States, and their firms face different domestic and export markets. Industrial policy, occasionally a dirty term in the United States, enjoys much more favor in Europe. Perhaps most important, the European Union has mechanisms for the domestic harmonization of law that have significant implications for how different countries treat IP.

The existing international intellectual property regime has come about through the collective action of a group of large corporations exerting their influence on national governments and on the multilateral rule creation process. These firms stood to profit from the extension of IP protection to the international stage, or stood to suffer significant losses from changes in technology that made IP piracy easier. While these firms were concentrated in sectors that have traditionally placed a strong emphasis on intellectual property (communications, pharmaceuticals, entertainment), this process has had unexpected and unpredictable side effects with respect to the development of IP regulation and protection in other fields. In particular, the emergence of crossover dual-use technology has made IP management and regulation a central concern of defense departments and defense firms.

In part, this response by defense-oriented corporations is the result of an increase in the potential of industrial espionage of defense technology. With respect to industrial espionage, a full analysis of the implications of cyber warfare for illicit acquisition of military technology requires considerably more research. This book hopes to lay the foundation for that research. The future of the Chinese military-industrial complex depends to some extent on its ability to acquire technology from the United States and elsewhere. Therefore, the future of US-Chinese relations depends to some extent on the size and sophistication of the Chinese defense industrial base. Given the importance of the US-China military balance for twenty-first-century geopolitics, the determinants of China's military innovation capacity merit greater attention.

Perhaps most important, this book has demonstrated that states and military organizations take international and domestic legal structures into account when making decisions about national security. These legal concerns and considerations go to the core of state policymaking in the defense sphere, setting the terms by which national governments make decisions about how to protect themselves and their property, both physical and intellectual. Contrary to the expectations of realist scholarship, defense ministries take legal constraints seriously, even when faced with a dangerous, anarchic international system.

Even under anarchy, states invariably pursue multiple goals, weighing the importance of military security against such issues as economic prosperity, international reputation, and the protection of domestic producers, which are themselves guarantees of a state's position in the global hierarchy. Defense ministries, in most cases, must therefore negotiate with other stakeholders while pursuing the goals that they seek. Consequently, professional military organizations often find themselves constrained by law, especially when domestic bureaucracies for managing particular legal questions have developed.

The extent to which states accept legal constraints on their ability to provide for the common defense — whether in the field of IP, the field of the law of armed conflict, or other areas — remains an important subject of study within the field of international relations. Indeed, it goes to the core of the question of how the institutionalization of international politics in the postwar world has affected the prospects for and conduct of conflict — not to mention how globalization and the resultant economic interconnectedness of states and corporations has affected the prospects for interstate conflict.

Conclusion

The study and practice of international relations has long stood in tension with the study and practice of international law. Scholars of the former have often seen efforts associated with the latter as a naive and unnecessary invasion of the former. E. H. Carr, in particular, argued that the post–World War I efforts to establish a legal regime capable of managing international politics would founder upon the collision of interest and power among great nations.[30] For their part, advocates of institutionalization have sometimes approached "realist" visions of international politics as apologetics for the misbehavior of great states.

This tension has made it difficult to assess the real impact of domestic and international law on how states defend themselves, supposedly a core interest of international relations scholars. Hopefully, this book has demonstrated that we cannot understand how states develop and acquire weapons without appreciating the domestic and international legal frameworks under which they operate. States compete for their security, but they compete within social and legal frameworks that channel and focus their efforts. These frameworks are, in and of themselves, subject to the competition of states, as power and interest affect the construction and functioning of institutions.

ACKNOWLEDGMENTS

We would like to thank the research assistants whose work has made this book possible. These include Catherine Putz, Tyler Lovell, Julian Fischer-Lhamon, Patrick Smith, and Christina Zeidan. We are particularly grateful to Charles Dainoff and Erik Fay, who made the quantitative possible.

We are deeply in debt for advice and assistance from scholars and practitioners familiar with this field. These include James Hasik, Greg Sanders, Catherine Cabrera, Dan Nexon, Stephen Brooks, Michael Desch, and Eugene Gholz.

We are also indebted to the University of Kentucky, the United States Army War College, and especially the Patterson School of Diplomacy and International Commerce for support during the writing of this book. In particular, Dr. Kathleen Montgomery, Ambassador Carey Cavanaugh, and Dr. Greg Hall have been of tremendous help during this project.

Finally, we would like to thank Charles Myers, who has adeptly husbanded this project at the University of Chicago Press. We are forever grateful to the late Christopher Rhodes, who gave the initial go-ahead to this book.

NOTES

Chapter 1

1.   C. J. Chivers, *The Gun* (Simon & Schuster, 2010), 164.
2.   Aaron Karp, "The Global Small Arms Industry: Transformed by War and Society," in Richard Bitzinger. ed., *The Modern Defense Industry* (ABC Clio, 2009), 274.
3.   Definitive estimates of the number of AK-47s produced, or even of the total number of producing states, are extremely difficult to find. *Small Arms Survey* suggests that as many as 100 million or more AK-47 variants have been produced worldwide. "Continuity and Change: Products and Producers," *Small Arms Survey 2004* (Graduate Institute of International and Development Studies, 2004). http://www.smallarmssurvey.org/fileadmin/docs/A-Yearbook/2004/en/Small-Arms-Survey-2004-Chapter-01-EN.pdf.
4.   Chivers, *The Gun*, 184.
5.   Chivers, *The Gun*, 190.
6.   Chivers, *The Gun*, 194.
7.   Stephen G. Brooks, *Producing Security: Multinational Corporations, Globalization, and the Changing Calculus of Conflict* (Princeton University Press, 2007).
8.   *Small Arms Survey 2007* (Graduate Institute of International and Development Studies, 2007), 24.
9.   Chivers, *The Gun*, 259.
10. Chivers, *The Gun*, 260.
11. Chivers, *The Gun*, 279.
12. Chivers, *The Gun*, 277.
13. Chivers, *The Gun*, 275.
14. Chivers, *The Gun*, 278.
15. Chivers, *The Gun*, 280.
16. Chivers, *The Gun*, 297.
17. "Military Assault Rifles," *Small Arms Survey 2013* (Graduate Institute of International and Development Studies, 2013), 1. http://www.smallarmssurvey.org/fileadmin/docs/H-Research_Notes/SAS-Research-Note-25.pdf.
18. Karp, "Global Small Arms Industry," 275.
19. *Small Arms Survey Yearbook 2007: Guns and the City* (Cambridge University Press, 2007), 19.
20. Robert L. Paarlberg, "Knowledge as Power: Science, Military Dominance, and US Security." International Security 29, no. 1 (2004): 135.

21. Susan K. Sell, *Private Power, Public Law: The Globalization of Intellectual Property Rights*, Kindle edition (Cambridge University Press, 2003).

22. James M. Hasik, *Arms and Innovation: Entrepreneurship and Alliances in the Twenty-First-Century Defense Industry* (University of Chicago, 2008).

23. Daniel Fiott, "Europe and the Pentagon's Third Offset Strategy," *RUSI Journal* 161, no. 1 (2016): 26–31, 27.

24. Mario Daniels, "Restricting the Transnational Movement of 'Knowledgeable Bodies': The Interplay of US Visa Restrictions and Export Controls in the Cold War," in John Krige, ed., *How Knowledge Moves: Writing the Transnational History of Science and Technology*, (University of Chicago Press, 2019), 40.

25. Chuck Hagel, "Secretary of Defense Speech: Reagan National Defense Forum Keynote," US Department of Defense, November 15, 2014, http://www.defense.gov /News/Speeches/Speech-View/Article/606635. See also Katherine Hicks et al., "Assessing the Third Offset Strategy," Center for Strategic and International Studies, March 2017, https://csis-prod.s3.amazonaws.com/s3fs-public/publica tion/170302_Ellman_ThirdOffsetStrategySummary_Web.pdf.

26. See, for example, Scott Sagan, *The Limits of Safety: Organizations, Accidents, and Nuclear Weapons* (Princeton University Press, 1993); Kimberley Martin Zisk, *Engaging the Enemy: Organization Theory and Soviet Military Innovation, 1955–1991* (Princeton University Press, 1993); Barry Posen, *The Sources of Military Doctrine: France, Britain, and Germany between the World Wars* (Cornell University Press, 1984); Elizabeth Kier, "Culture and Military Doctrine: France between the Wars," *International Security* 19, no. 4 (Spring 1995): 69–70; Mary Habeck, *Storm of Steel: The Development of Armor Doctrine in Germany and the Soviet Union, 1919–1939* (Cornell University Press, 2003); Elizabeth Kier, *Imagining War: French and British Military Doctrine between the Wars* (Princeton University Press, 1997).

27. See, for example, Williamson Murray and Allan R. Millett, eds., *Military Innovation in the Interwar Period* (Cambridge University Press, 1996); Allan R. Millet and Williamson Murray, eds., *Military Effectiveness*, vol. 1 (Allen and Unwin, 1988); Deborah D. Avant, *Political Institutions and Military Change: Lessons from Peripheral Wars* (Cornell University Press, 1994).

28. See, for example, Stephen Peter Rosen, *Societies and Military Power: India and its Armies* (Cornell University Press, 1996).

29. Stephen Peter Rosen, *Winning the Next War: Innovation and the Modern Military* (Cornell University Press, 1991).

30. Colin Gray, *Weapons Don't Make War: Policy, Strategy, and Military Technology* (University Press of Kansas, 1993).

31. See, for example, Keir Leiber, *War and the Engineers: The Primacy of Politics over Technology* (Cornell University Press, 2005).

32. Mike Pietrucha, "The Search for the Technological Silver Bullet to Win Wars," War on the Rocks, August 26, 2015. https://warontherocks.com/2015/08/the-search -for-the-technological-silver-bullet-to-win-wars/.

33. George H. Quester, *Offense and Defense in the International System* (Wiley, 1977).

34. Peter Gray, *Leadership, Direction and Legitimacy of the RAF Bomber Offensive from Inception to 1945* (Continuum, 2012), 115.

35. Robert Jervis, "Cooperation under the Security Dilemma," *World Politics* 30, no. 2 (January 1978): 167–214. See also Stephen Van Evera, "Offense, Defense, and the Causes of War," *International Security* 22, no. 4 (Spring 1998): 5–43; Charles L. Glaser and Chaim Kaufmann. "What Is the Offense-Defense Balance and Can We Measure It?" *International Security* 22, no. 4 (Spring 1998): 44–82; Sean M. Lynn-Jones, "Offense-Defense Theory and Its Critics," *Security Studies* 4, no. 4 (Summer 1995): 660–91.

36. Keir A. Lieber, "Grasping the Technological Peace: The Offense-Defense Balance and International Security," *International Security* 25, no. 1 (Summer 2000): 71–104.

37. Stephen Biddle, *Military Power: Explaining Victory and Defeat in Modern Battle* (Princeton University Press, 2004).

38. William D. Nordhaus, "Theory of Innovation: An Economic Theory of Technological Change," *American Economic Review* 59, no. 2 (May 1969): 18.

39. Gray, *Weapons Don't Make War*, 116.

40. Robert Neild, "Defining 'Offensive': A Failure and a Success," *Bulletin of Atomic Scientists* (September 1988), 18.

41. Gray, *Weapons Don't Make War*, 120.

42. Ian Bellany, "The Offensive-Defensive Distinction, the International Arms Trade, and Richardson and Dewey," *Peace and Conflict: Journal of Peace Psychology* 1, no. 1 (1995): 37–48.

43. Andrew F. Krepinevich, "Cavalry to Computer: The Pattern of Military Revolutions," *National Interest* 37 (Fall 1994): 30–42.

44. Barry D. Watts, *The Evolution of Precision Strike* (Center for Strategic and Budgetary Assessments, 2013), 5.

45. For a discussion of the role of ideational versus technological factors in the emergence of mechanized warfare, see Michael Geyer, "German Strategy in the Age of Machine Warfare, 1914–1945," in Peter Paret and Gordon Craig, eds., *Makers of Modern Strategy: From Machiavelli to the Nuclear Age* (Princeton University Press, 1986), 556–57.

46. Robert R. Tomes, "Relearning Counterinsurgency Warfare," *Parameters* (Spring 2004): 16–28; 17.

47. Michael C. Desch, "Don't Worship at the Altar of Andrew Marshall," *National Interest*, December 17, 2014, http://nationalinterest.org/feature/the-church-st-andy-11867.

48. Dima Adamsky, *The Culture of National Innovation: The Impact of Cultural Factors on the Revolution in Military Affairs in Russia, the US, and Israel* (Stanford University Press, 2010).

49. Barry Watts, "Precision Strike: An Evolution," *National Interest*, November 2, 2013, http://nationalinterest.org/commentary/precision-strike-evolution-9347.

50. Paarlberg, "Knowledge as Power," 135.

51. Andrew Krepinevich and Barry Watts, *The Last Warrior: Andrew Marshall and the Shaping of Modern American Defense Strategy* (Basic Books, 2015).

52. Tai Ming Cheung, Thomas G. Mahnken, and Andrew L. Ross, "Frameworks for Analyzing Chinese Defense and Military Innovation," in Tai Ming Cheung, ed.,

Forging China's Military Might: A New Framework for Assessing Innovation (Johns Hopkins University Press, 2014). See also Charles Edquist, *Systems of Innovation: Technologies, Institutions, and Organizations* (Pinter, 1997); Bengt-Ake Lundvall, ed., *National Systems of Innovation: Towards a Theory of Innovation and Interactive Learning* (Pinter, 1992); Richard R. Nelson, ed., *National Innovation Systems: A Comparative Analysis* (Oxford University Press, 1993).

53. Paul Bracken, Linda Brandt, and Stuart E. Johnson, "The Changing Landscape of Defense Innovation," *Defense Horizons* 47 (July 2005): 3.

54. Tai Ming Cheung, "An Uncertain Tradition," in Tai Ming Cheung, ed., *Forging China's Military Might: A New Framework for Assessing Innovation*, Kindle edition (Johns Hopkins University Press, 2014), 1006.

55. Cheung, "An Uncertain Tradition," 1029.

56. Emily O. Goldman and Leslie C. Eliason, *The Diffusion of Military Technology and Ideas* (Stanford University Press, 2003).

57. Kenneth Waltz, *Theory of International Politics* (McGraw-Hill, 1979), 127.

58. Waltz, *Theory*, 127.

59. George P. Huber, "Organizational Learning: The Contributing Processes and Literatures," *Organization Science* 2, no. 1 (1991): 91.

60. Huber, "Organizational Learning," 96.

61. Paul J. DiMaggio and Walter W. Powell, "The Iron Cage Revisited: Institutional Isomorphism and Collective Rationality in Organizational Fields," *American Sociological Review* 28, no. 2 (April 1983): 150.

62. Barry Posen, *The Sources of Military Doctrine: France, Britain, and Germany between the World Wars* (Cornell University Press, 1984).

63. João Resende-Santos, *Neorealism, States, and the Modern Mass Army*, Kindle edition (Cambridge University Press, 2007).

64. Michael Horowitz, *The Diffusion of Military Power: Causes and Consequences for International Politics* (Princeton University Press, 2010).

65. Andrea Gilli and Mauro Gilli, "The Diffusion of Drone Warfare? Industrial, Organizational, and Infrastructural Constraints," *Security Studies* 25, no. 1: 50–84.

66. Andrea Gilli and Mauro Gilli. "Military-Technological Superiority, Systems Integration and the Challenges of Imitation, Reverse Engineering, and Cyber-Espionage." *International Security* (2018).

67. Douglas M. O'Reagan, *Taking Nazi Technology: Allied Exploitation of German Science after the Second World War*, Kindle edition (Johns Hopkins University Press, 2019), 371.

68. John W. Meyer, John Boli, George M. Thomas, and Francisco O. Ramirez. "World Society and the Nation-State." *American Journal of Sociology* 103, no. 1 (July 1997): 144–81; 146.

69. Martha Finnemore, *The Purpose of Intervention: Changing Beliefs about the Use of Force* (Cornell University Press, 2003), 16.

70. Emily Goldman, "The Spread of Western Military Models to Ottoman Turkey and Meiji Japan," in Theo Farrell and Terry Terriff, eds., *The Sources of Military Change: Culture, Politics, Technology* (Lynne Riener, 2002), 41–68, 43; Theo Farrell, "World

Culture and the Irish Army, 1922–1942," in Theo Farrell and Terry Terriff, eds., *The Sources of Military Change: Culture, Politics, Technology*, 69–90; 82.

71. Dana P. Eyer and Mark C. Suchman, "Status, Norms, and the Proliferation of Conventional Weapons: An Institutional Theory Approach," in Peter J. Katzenstein, ed., *The Culture of National Security* (Columbia University Press, 1996), 81.

72. Daniel W. Henk and Marin Revavi Rupiya, *Funding Defense: Challenges of Buying Military Capability in Sub-Saharan Africa* (Strategic Studies Institute, 2001), 20.

73. Amitai Etzioni, "International Prestige, Competition and Peaceful Coexistence," *European Journal of Sociology* 3, no. 1 (1962): 21–41.

74. Dennis M. Gormley, *Missile Contagion: Cruise Missile Proliferation and the Threat to International Security* (Naval Institute Press, 2008); 6.

75. Gormley, *Missile Contagion*, 108.

76. Gormley, *Missile Contagion*, 124.

77. Marcus Tullius Cicero, *Cicero pro milone* (Macmillan, 1909), available at http://www.thelatinlibrary.com/cicero/milo.shtml. Via Stephanie Carvin and Michael John Williams, *Law, Science, Liberalism, and the American Way of Warfare*, Kindle edition (Cambridge University Press, 2014), 987.

78. Carl von Clausewitz, *On War*, Michael Howard and Peter Paret, trans. (Everyman, 1993), 710.

79. See, for example, E. H. Carr, *The Twenty Years' Crisis: 1919–1939* (Perennial, 2001 [1939]).

80. Peter Gourevitch, "The Second Image Reversed: The International Sources of Domestic Politics." *International Organization* 32, no. 4 (1978), 881–912. http://www.jstor.org/stable/2706180.

81. Kenneth Waltz, *Man, the State, and War: A Theoretical Analysis* (Columbia University Press, 1959), 80.

82. See, for example, Martha Finnemore and Kathryn Sikkink, "International Norm Dynamics and Political Change," *International Organization* 52, no. 4 (1998): 887–917; Geoffrey Garrett, "Global Markets and National Politics: Collision Course or Virtuous Circle?" *International Organization* 52, no. 4 (1998): 787–824; Randall L. Schweller, "Domestic Structure and Preventive War: Are Democracies More Pacific?" *World Politics* 44, no. 2 (1992): 235–69; Ronald Rogowski, *Commerce and Coalitions: How Trade Affects Domestic Political Alignments* (Princeton University Press, 1989).

83. Edward C. Luck, "Gaps, Commitments, and the Compliance Challenge," in Edward C. Luck and Michael W. Doyle, eds., *International Law and Organization: Closing the Compliance Gap* (Rowman and Littlefield, 2004). 307.

84. Abram Chayes and Antonia Handler Chayes. "On Compliance," *International Organization* 47, no. 02 (1993): 175–205.

85. Harold Hongju Koh, Abram Chayes, Antonia Handler Chayes, and Thomas M. Franck, "Why Do Nations Obey International Law?" *Yale Law Journal* 106, no. 8 (1997): 2599–2659.

86. Katharina P. Coleman and Michael W. Doyle, "Introduction: Expanding Norms,

Lagging Compliance," in Luck and Doyle, eds., *International Law and Organization: Closing the Compliance Gap*, 7.

87. Coleman and Doyle, "Introduction," 7.

88. Legal Information Institute, "Self-Executing Treaty," Cornell University Law School, https://www.law.cornell.edu/wex/self_executing_treaty.

89. Jonas Tallberg, "Paths to Compliance: Enforcement, Management, and the European Union," *International Organization* 56, no. 3 (2002): 609–43.

90. Ernst B. Haas, *Beyond the Nation-State: Functionalism and International Organization* (Stanford University Press, 1964).

91. Alec S. Sweet, Wayne Sandholtz, and Neil Fligstein, *The Institutionalization of Europe* (Oxford University Press, 2001).

92. Judith Goldstein, "International Law and Domestic Institutions: Reconciling North American "Unfair" Trade Laws." *International Organization* 50, no. 4 (1996): 541–64; 562.

93. Steven E. Lobell, "Second Image Reversed Politics: Britain's Choice of Freer Trade or Imperial Preferences, 1903–1906, 1917–1923, 1930–1932." *International Studies Quarterly* 43, no. 4 (1999): 671–93.

94. Gary D. Solis, *The Law of Armed Conflict: International Humanitarian Law in War* (Cambridge University Press, 2010); Andrew Clapham, *Human Rights Obligations of Non-State Actors* (Oxford University Press, 2006).

95. Harold A. Feiveson and Jacqueline W. Shire, "Dilemmas of Compliance with Arms Control and Disarmament Agreements," in Luck and Doyle, eds., *International Law and Organization: Closing the Compliance Gap*, 209.

96. Michael Patrick Ryan, *Knowledge Diplomacy: Global Competition and the Politics of Intellectual Property* (Brookings Institution Press, 1998), 16.

97. Catherine Saez, "Review of WIPO Technical Assistance, Four Years after Release, Still Stirs Up Development Committee," *Intellectual Property Watch*, December 11, 2015, http://www.ip-watch.org/2015/11/12/review-of-wipo-technical-assistance -four-years-after-release-still-stirs-up-development-committee/.

98. Henry Farrell and Abraham Newman. "Weaponized Interdependence," *International Security* 44, no. 1 (2019).

99. Sell, *Private Power, Public Law*, 96.

100. Daniel W. Drezner, "Globalization and Policy Convergence." *International Studies Review* 3, no. 1 (2001): 53–78.

101. Waltz, *Theory*, 127; Horowitz, *Diffusion of Military Power*, 21.

Chapter 2

1. Adrian Johns, *Piracy: The Intellectual Property Wars from Gutenberg to Gates*, Kindle edition (University of Chicago Press, 2010), 974.

2. Johns, *Piracy*, 1034.

3. Johns, *Piracy*, 1038.

4. Johns, *Piracy*. 1038.

5. Johns, *Piracy*, 974.

6. Jeremy Black, *Naval Warfare: A Global History since 1860* (Rowman & Littlefield, 2017), 4.
7. Black, *Naval Warfare*, 5.
8. Johns, *Piracy*, 3453.
9. Johns, *Piracy*, 3557.
10. Black, *Naval Warfare*, 33–36.
11. Katherine C. Epstein, *Torpedo* (Harvard University Press, 2014), 211.
12. Epstein, *Torpedo*, 4.
13. Epstein, *Torpedo*, 4.
14. Epstein, *Torpedo*, 74.
15. Epstein, *Torpedo*, 74.
16. Epstein, *Torpedo*, 75.
17. The fourth form of intellectual property, trademarks, are "source identifiers"— casually understood as brand names or logos. Source identifiers achieve several goals, including signaling to consumers (including governments) that a particular product will have positive characteristics—such as design and production quality—that cannot be confirmed prior to purchase. Trademarks are not particularly relevant to the defense industry discussions in this book, for several reasons. First, as only source identifiers, trademarks are not relevant to either assuring or preventing the dissemination of confidential information. Second, as discussed further in this book, governments have tended to procure their products from relatively few suppliers, whose quality they track (even if that means weakening intellectual property enforcement to do so). As a result, source identifiers play a much smaller role in affecting defense purchasing decisions than in the open market. Finally, as many defense products are acquired only through complex defense acquisition processes, which contain specific design and production specifications, those design and production quality assurances created by trademarks are far less significant.
18. To receive a utility patent, an applicant must disclose detailed mechanics of his invention, and explain not only how the technology differs from prior creations in the field but how it is sufficiently original so that the common technician in the field could not have developed it on her own.
19. "Vault of the Secret Formula," World of Coca-Cola, http://www.worldofcoca-cola .com/explore/explore-inside/explore-vault-secret-formula/.
20. Jack Percher, "Samsung Granted Design Patents for an In-Vehicle Dashboard Cluster, a Series of New Tablet User Interfaces & More," January 21, 2014, http:// www.patentlymobile.com/2014/01/samsung-granted-design-patents-for-an-in -vehicle-dashboard-cluster-a-series-of-new-tablet-user-interfaces-more.html; Joshua D. Wolson, "Opinion: When Form Trumps Function, and Common Sense: Problems with Damages for Design Patents," InsideSources, July 23, 2015, http:// www.insidesources.com/when-form-trumps-function-and-common-sense -problems-with-damages-for-design-patents/.
21. The US Supreme Court handed down an opinion in 2014 that many thought would strip patent protection from almost all software. However, the Federal Cir-

cuit Court, which is generally the final arbiter of patent lawsuits in the US courts, has started using an apparently narrow exception in the Supreme Court's decision to uphold the legitimacy of most software patents. Ben Klemens, "Software Patents Poised to Make a Comeback under New Patent Office Rules," Ars Technica, January 10, 2019. https://arstechnica.com/tech-policy/2019/01/software-patents-poised-to-make-a-comeback-under-new-patent-office-rules/. See also Lily Hay Newman, "Who Owns the Software in the Car You Bought?" Slate, May 22, 2015. http://www.slate.com/blogs/future_tense/2015/05/22/gm_and_john_deere_say_they_still_own_the_software_in_cars_customers_buy.html.

22. "Copyright Registration of Computer Programs," www.copyright.gov/circs/circ 61.pdf; David Kravets, "US Regulators Grant DMCA Exemption Legalizing Vehicle Software Tinkering," Ars Technica, October 27, 2015. http://arstechnica.com/tech-policy/2015/10/us-regulators-grant-dmca-exemption-legalizing-vehicle-software-tinkering/.

23. Johns, *Piracy*, 3477.

24. Johns, *Piracy*, 3231.

25. Johns, *Piracy*, 3687.

26. Johns, *Piracy*, 5100.

27. Johns, *Piracy*, 5192.

28. Johns, *Piracy*, 5201.

29. Johns, *Piracy*, 5214.

30. Johns, *Piracy*, 5280. See also Polanyi, Michael (1944), "Patent Reform," *Review of Economic Studies* 11 (1944): 61–76.

31. William D. Nordhouse, "Theory of Innovation: An Economic Theory of Technological Change," *American Economic Review* 59, no. 2 (May 1969): 18.

32. Jacob Schmooker, "Economic Sources of Inventive Activity," *Journal of Economic History* 22, no. 1 (1962): 1–20; 1.

33. Schmooker, "Economic Sources of Inventive Activity," 2.

34. Schmooker, "Economic Sources of Inventive Activity," 4.

35. William D. Nordhaus, "The Optimum Life of a Patent: Reply," *American Economic Review* 62, no. 3 (1972): 431.

36. Ugo Pagano, "The Crisis of Intellectual Monopoly Capitalism," *Cambridge Journal of Economics* 38, no. 6 (2014): 1409.

37. Mike Kimel, "Do Patents Lead to Economic Growth?" Angry Bear, March 27, 2017. https://angrybearblog.com/2017/03/do-patents-lead-to-economic-growth.html.

38. Bjørn L. Basberg, "Patents and the Measurement of Technological Change: A Survey of the Literature," *Research Policy* 16, no. 2 (1987): 131–41.

39. Edwin L.-C. Lai, "International Intellectual Property Rights Protection and the Rate of Product Innovation," *Journal of Development Economics* 55, no. 1 (1998): 133–53; Dan L. Burk and Mark A. Lemley, "Policy Levers in Patent Law," *Virginia Law Review* (2003): 1575–1696; Mariko Sakakibara and Lee Branstetter, "Do Stronger Patents Induce More Innovation? Evidence from the 1988 Japanese Patent Law Reforms," *NBER Working Papers Series*, no. 7066 (April 1999): 2; Wil-

liam D. Nordhaus, "An Economic Theory of Technological Change," *American Economic Review* 59, no. 2 (May 1969): 18–28.

40. Lai, "International Intellectual Property Rights Protection," 134.

41. Sakakibara and Branstetter, "Do Stronger Patents Induce More Innovation?" 2.

42. Sakakibara and Branstetter, "Do Stronger Patents Induce More Innovation?" 31.

43. Sakakibara and Branstetter, "Do Stronger Patents Induce More Innovation?" 31.

44. Bryan Kelly, Dimitris Papanikolaou, Amit Seru, and Matt Taddy, "Measuring Technological Innovation over the Long Run," National Bureau of Economic Research working paper no. 25266, 2018. https://www.nber.org/papers/w25266. Further concluding that spikes in patent quality successfully predicted watershed inventions as well as individual firm profits, suggesting that innovation might indeed be key to understanding the last two centuries of economic growth.

45. Basberg, "Patents and the Measurement of Technological Change," 132–34.

46. Dan L. Burk and Mark A. Lemley, "Is Patent Law Technology-Specific?" *Berkeley Technology Law Journal* 17 (2002): 1155–56.

47. Albert Guangzhou Hu and Ivan P. Png, "Patent Rights and Economic Growth: Evidence from Cross-Country Panels of Manufacturing Industries," CELS 2009 4th Annual Conference on Empirical Legal Studies paper, August 10, 2012. https://www.wipo.int/edocs/mdocs/mdocs/en/wipo_ip_econ_ge_5_10/wipo_ip_econ_ge_5_10_ref_huandpng.pdf.

48. Hu and Png, "Patent Rights and Economic Growth."

49. Michael A. Heller and Rebecca S. Eisenberg. "Can Patents Deter Innovation? The Anticommons in Biomedical Research." Science 280, no. 5364 (1998): 698.

50. James Pooley, "Trade Secrets: The Other Intellectual Property Right," WIPO Magazine (June 2013), https://www.wipo.int/wipo_magazine/en/2013/03/article_0001.html.

51. Andrea Fosfuri and Thomas Rønde, "High-Tech Clusters, Technology Spillovers, and Trade Secret Laws," *International Journal of Industrial Organization* 22, no. 1 (2004): 47–48.

52. See, for example, a Google's subsidiary's ability to enforce its trade secret protection in the US courts: David Pridham, "Steal a Trade Secret, Go to Jail?" *Forbes*, June 1, 2017. https://www.forbes.com/sites/davidpridham/2017/06/01/steal-a-trade-secret-go-to-jail/#683fcbaa6a47.

53. See Michael A. Riley and Ashlee Vance, "China Corporate Espionage Boom Knocks Wind Out of U.S. Companies," *Bloomberg*, March 15, 2012. http://www.bloomberg.com/news/articles/2012-03-15/china-corporate-espionage-boom-knocks-wind-out-of-u-s-companies.

54. Discussed later in great detail is China's unwillingness to respect attempts to protect trade secrets: Christopher Burgess, "China Continues to Steal High-Tech Trade Secrets," CSO, May 27, 2017. https://www.csoonline.com/article/3198664/security/china-continues-to-steal-high-tech-trade-secrets.html.

55. Susan K. Sell, *Private Power, Public Law: The Globalization of Intellectual Property Rights* (Cambridge University Press, 2003).

56. Susan K. Sell, "Intellectual Property Rights in Historical Perspective," *Loyola of*

Los Angeles Law Review (September 1, 2004): 60–74, noting that, "since the four-teenth and fifteenth centuries, this policy often focused on technology transfer and diffusion."; P. J. Federico, "The Origin of Patents," *Journal of the Patent Office Society* 11 (1929): 292, 293.

57. Michael P. Ryan, *Knowledge Diplomacy: Global Competition and the Politics of Intel-lectual Property* (Brookings Institution Press, 1998), 26.

58. Joseph Nye, "Globalism Versus Globalization," April 15, 2002, https://www.the globalist.com/globalism-versus-globalization. Nye notes that economic global-ism rose between 1850 and 1914—and fell between 1914 and 1945. See also the 1672 Mercantilist Rule in Colony of Connecticut: "There shall be no monopo-lies granted or allowed amongst us but of such new inventions as *shall be judged profitable for the country* and for such time as the general court shall judge meet." (Emphasis added.) Thomas Jefferson, letter to Isaac McPherson, August 13, 1813: "Society may give an exclusive right to the profits arising from them, as an en-couragement to men to pursue ideas which may produce utility, but this may or may not be done according to the will and convenience of the society, without claim or complaint from anybody."

59. Peter Drahos and John Braithwaite, *Information Feudalism: Who Owns the Knowl-edge Economy?* (The New Press, 2007), 32.

60. Sell, "Intellectual Property Rights in Historical Perspective."

61. Sell, "Intellectual Property Rights in Historical Perspective."

62. Peter Drahos and John Braithewaite, "Who Owns the Knowledge Economy? Po-litical Organising behind TRIPS," Corner House briefing no. 32 (September 2004), 32, http://www.thecornerhouse.org.uk/item.shtml?x=85821; Robert Merges, "Battle of the Lateralisms: Intellectual Property and Trade," *Boston University International Law Journal* 8, no. 2 (Fall 1990): 245.

63. Merges, "Battle of the Lateralisms." The same held true for copyright. During the 1830s, the US publishing industry engaged in widespread piracy of English liter-ary output, leading numerous British authors—most famously, an angry Charles Dickens—to petition Congress in 1836 for protection for non-US authors. Such requests were wholly unsuccessful.

64. Douglas M. O'Reagan, *Taking Nazi Technology: Allied Exploitation of German Sci-ence after the Second World War*, Kindle edition (Johns Hopkins University Press, 2019), 1097–1100.

65. Joshua Aizenman and Reuven Glick suggest that, particularly in low-income countries, the presence of military conflict may motivate countries to expand intellectual property protections in order to encourage development of defense technology. Joshua Aizenman and Reuven Glick, "Military Expenditure, Threats, and Growth," *Journal of International Trade & Economic Development* 15, no. 2 (2006): 129–55.

66. Johns, *"Piracy,"* 3969.

67. Johns, *"Piracy,"* 3981.

68. Frank Thadeusz, "No Copyright Law: The Real Reason for Germany's Industrial Expansion," *Der Spiegel*, August 18, 2010.

69. Johns, *"Piracy,"* 3435.

70. Sell, "Intellectual Property Rights in Historical Perspective."
71. Ryan, *Knowledge Diplomacy*, 50.
72. "TRIPS: Agreement on Trade-Related Aspects of Intellectual Property Rights," Apr. 15, 1994, Marrakesh Agreement Establishing the World Trade Organization, Annex 1C, 1869 U.N.T.S. 299, 33 I.L.M. 1197.
73. Drahos and Braithewaite, "Who Owns the Knowledge Economy?" 38 (emphasis added).
74. A fairly small number of (generally poorer) states have entirely outsourced their patenting authority, relying on international regional patent offices to handle their patenting. Jeffrey Shieh, "An Overview of the PCT International Patent Process," IPWatchdog.com, June 29, 2013. http://www.ipwatchdog.com/2011/08/18/an-overview-of-the-pct-international-patent-process/id=18805/; "PCT Contracting States for Which a Regional Patent Can Be Obtained via the PCT," WIPO: World Intellectual Property Organization, November 1, 2015. http://www.wipo.int/pct/en/texts/reg_des.html.
75. Timothy R. Holbrook, "Extraterritoriality in U.S. Patent Law," *William & Mary Law Review* 49, no. 6 (2008). Multiple countries' patents can also be litigated in private arbitration when there has been a prior private agreement between parties to do so. Felicia Boyd, "A Way to Efficiently Resolve International Patent Disputes," Law360: The Newswire for Business Lawyers, February 28, 2017. http://www.law360.com/articles/743379/a-way-to-efficiently-resolve-international-patent-disputes;
76. "Huawei Has Been Cut Off from American Technology," *Economist*, May 25, 2019. https://www.economist.com/business/2019/05/25/huawei-has-been-cut-off-from-american-technology.
77. There is additional administrative coordination between some state patent agencies. The European Patent Office coordinates with the Japan Patent Office, the Korean Intellectual Property Office, the State Intellectual Property Office of the People's Republic of China, and the US Patent and Trademark Office to develop efficiency in patent application processing. The US Patent Office's description of the "IP5" notes that those five offices account for 90 percent of all patent applications filed worldwide. Office of Policy and International Affairs, "Office of Policy and International Affairs: IP5," http://www.uspto.gov/patents-getting-started/international-protection/office-policy-and-international-affairs-ip5.
78. Drahos and Braithwaite, "Who Owns the Knowledge Economy?"
79. The proliferation of so-called "TRIPS-plus" bilateral accords that impose minimum standards for IP protection are even *more* stringent than those of TRIPS.
80. Ryan, *Knowledge Diplomacy*, 68.
81. Drahos and Braithewaite, "Who Owns the Knowledge Economy?"
82. *Berne Convention for the Protection of Literary and Artistic Works, of September 9, 1886, Completed at Paris on May 4, 1896, Revised at Berlin on November 13, 1908, Completed at Berne on March 20, 1914, Revised at Rome on June 2, 1928, Revised at Brussels on June 26, 1948, and Revised at Stockholm on July 14, 1967*. 1967. United International Bureau for the Protection of Intellectual Property, 1967.
83. Sell, "Intellectual Property Rights in Historical Perspective."

84. Steven Seidenberg, "TPP Strengthens Controversial IP Arbitration," Intellectual Property Watch, November 30, 2015. http://www.ip-watch.org/2015/11/30/tpp-strengthens-controversial-ip-arbitration/.

85. The original version was called the Trans-Pacific Partnership. The agreement could not enter into force once the United States withdrew its signature. In January 2018, US President Donald Trump in an interview announced his interest in possibly rejoining the TPP if it were a "substantially better deal" for the United States. The developing Transatlantic Trade and Investment Partnership, between the European Union and the United States, contained a similar provision; the negotiations of that expected agreement were halted indefinitely following the 2016 United States presidential election. By mid-2017, representatives of both the United States and the European Union expressed willingness to resume the negotiations.

86. "TTIP and TPP under Dcrutiny: How to Assess the Intellectual Property Chapter." IP & IT at Your Fingertips, http://www.ipdigit.eu/2015/10/ttip-and-tpp-under-scrutiny-how-to-assess-the-intellectual-property-chapter.

87. Henrik Horn, Petros C. Mavroidis, and Andre Sapir, "Beyond the WTO? An Anatomy of EU and US Preferential Trade Agreements," CEPR discussion paper no. DP7317, June 2009, https://ssrn.com/abstract=1433913; Kaitlin Mara, "Stronger IP Enforcement Finds a Home in Bilateral Trade Agreements," Intellectual Property Watch, April 21, 2009, http://www.ip-watch.org/2009/04/21/stronger-ip-enforcement-finds-home-in-bilateral-trade-agreements/.

88. Office of the US Trade Representative, "Intellectual Property Rights in U.S.–South Korea Trade Agreement," https://ustr.gov/uskoreaFTA/IPR.

89. Office of the US Trade Representative, "Chile Free Trade Agreement," https://ustr.gov/trade-agreements/free-trade-agreements/chile-fta.

90. Mara, "Stronger IP Enforcement."

91. Josef Drexl, "Intellectual Property and Implementation of Recent Bilateral Trade Agreements in the EU," in *EU Bilateral Trade Agreements and Intellectual Property: For Better or Worse?* (Springer Berlin Heidelberg, 2014), 265–91.

Chapter 3

1. Davida H. Isaacs and Robert M. Farley, "Privilege-Wise and Patent (and Trade Secret) Foolish? How the Courts' Misapplication of the Military and State Secrets Privilege Violates the Constitution and Endangers National Security," *Berkeley Technology Law Journal* 24, no. 2 (2009): 785–818.

2. Sherry Sontag, Christopher Drew, and Annette Lawrence Drew, *Blind Man's Bluff: The Untold Story of Cold War Submarine Espionage* (Random House, 2000), 158.

3. *Crater Corp. v. Lucent Technologies, Inc.*, 625 F. Supp. 2d 790 (E.D. Mo. 2007).

4. *Crater Corp. v. Lucent Technologies, Inc.*, 625 F. Supp. 2d 790 (E.D. Mo. 2007).

5. *Crater Corp. v. Lucent Technologies, Inc.*, no. 4:98CV00913 ERW, 1999 WL 33973795 (E.D. Mo. August 25, 1999) (citing 28 U.S.C. § 1498(a)). See also the appellate decision: *Crater Corp. v. Lucent Technologies, Inc.*, 255 F.3d 1361, 1364 (Fed. Cir. 2001).

6. Daniel Larson, "Yesterday's Technology, Tomorrow: How the Government's Treat-

ment of Intellectual Property Prevents Soldiers from Receiving the Best Tools to Complete Their Mission," *J. Marshall Rev. Intell. Prop. L.* 7 (2007), 173.

7. Katherine C. Epstein, *Torpedo* (Harvard University Press, 2014), 15.

8. Edward A. Purcell, "Understanding Curtiss-Wright," *Law and History Review* 31, no. 4 (2013): 653–715; 657.

9. Epstein, *Torpedo*, 74–75.

10. See, for example, Abraham Rabinovich, *The Yom Kippur War: The Epic Encounter That Transformed the Middle East* (Schocken, 2007).

11. Barry Watts, *The Evolution of Precision Strike* (Center for Strategic and Budgetary Assessments, 2013), 5.

12. Watts, *Precision Strike*, 5.

13. Watts, *Precision Strike*, 6.

14. Watts, *Precision Strike*, 8.

15. Watts, *Precision Strike*, 8.

16. Watts, *Precision Strike*, 7–8.

17. Martin H. Weik, "The ENIAC Story," Ordnance Ballistic Research Laboratories, Aberdeen Proving Ground, 1961. https://ftp.arl.army/mil/~mike/comphist/eniac -story.html.

18. Weik, "The ENIAC Story."

19. John A. Alic, *Beyond Spinoff: Military and Commercial Technologies in a Changing World* (Harvard Business Press, 1992), 6. See also Sharon Weinberger, *The Imagineers of War: The Untold History of DARPA, the Pentagon Agency That Changed the World* (Alfred A. Knopf, 2017), 115–18.

20. Alic, *Beyond Spinoff*, 6.

21. Peter J. Dombroswki and Eugene Gholz. *Buying Military Transformation: Technological Innovation and the Defense Industry* (Columbia University Press, 2006), 5–7.

22. Watts, *Precision Strike*, 8.

23. Robert M. Farley, *Grounded: The Case for Abolishing the United States Air Force* (University Press of Kentucky, 2014).

24. Dombrowski and Gholz, *Buying Military Transformation*, 138.

25. James Hasik, *Arms and Innovation: Entrepreneurship and Alliances in the Twenty-First Century Defense Industry*, Kindle edition (University of Chicago Press, 2008), 52.

26. Dombrowski and Gholz, *Buying Military Transformation*, 138.

27. Dombrowski and Gholz, *Buying Military Transformation*, 139.

28. John Deutsch, "Consolidation of the US Defense Industrial Base," *Acquisition Review Quarterly* 8, no. 3 (2001): 138–50; 138.

29. Peter Dombrowski and Andrew Ross, "The Revolution in Military Affairs, Transformation, and the U.S. Defense Industry," in Richard A. Bitzinger, ed., *The Modern Defense Industry: Political, Economic, and Technological Issues: Political, Economic, and Technological Issues* (ABC-CLIO, 2009), 161.

30. Philip Finnegan, "The Evolution of Defense Hierarchies," in *The Modern Defense Industry*, 97.

31. Dombrowski and Ross, "Revolution," 161.

32. Jacques S. Gansler, William S. Greenwalt, William Lucyshin, *Non-Traditional Com-*

mercial Defense Contractors (Center for Public Policy and Private Enterprise, 2013), 12–13.

33. Gansler, *Commercial Defense*, v–vi.
34. Gansler, *Commercial Defense*, v–vi.
35. Hasik, *Arms and Innovation*, 74.
36. Hasik, *Arms and Innovation*, 113.
37. Finnegan, "Defense Hierarchies," 97.
38. Finnegan, "Defense Hierarchies," 98.
39. Hasik, *Arms and Innovation*, 116.
40. Julian Barnes, "Lockheed Lobbies for F-22 Production on Job Grounds," *Los Angeles Times*, February 11, 2009, http://articles.latimes.com/2009/feb/11/business/fi-jets11.
41. Finnegan, "Defense Hierarchies," 97.
42. Hasik, *Arms and Innovation*, 24.
43. Larson, "Yesterday's Technology," 179.
44. There are rare exceptions to this, primarily in the cases of contracts for construction work or architect-engineer services that involve only "standard types of construction"—a term defined in the Federal Acquisition Regulations.
45. Larson, "Yesterday's Technology," 184.
46. Interview with aerospace executive, April 20, 2016.
47. Gansler, Greenwalt, and Lucyshya, *Commercial Defense*, 53.
48. Interview with aerospace executive, April 20, 2016.
49. Bruce D. Jette, "Army Enacts New Policy on Intellectual Property," US Army Training and Doctrine Command, January 30, 2019. https://www.tradoc.army.mil/Publications-and-Resources/Article-Display/Article/1741504/army-enacts-new-policy-on-intellectual-property/.
50. Jon Harper, "Army Revamping Intellectual Property Policies," *National Defense*, August 21, 2018. http://www.nationaldefensemagazine.org/articles/2018/8/21/army-revamping-intellectual-property-policies.
51. Tom Ewing, "Five World War I Patents: War Tech Roundup." A New Domain, January 3, 2015, http://anewdomain.net/world-war-i-patents/.
52. Stephen, Van Dulken, "A Secret Aviation Invention from World War I," The Patent Search Blog, December 19, 2013, http://stephenvandulken.blogspot.com/2013/12/a-secret-aviation-invention-from-world.html.
53. Elizabeth Cregan, *The Effect of World War I on Radio*, http://www.academia.edu/937415/The_Impact_of_WWI_on_the_Course_of_American_Radio_History (2011), citing Susan J. Douglas, *Inventing American Broadcasting 1899–1922* (Johns Hopkins University Press, 1987), 102–43; and Paul Starr, *The Creation of the Media: Political Origins of Modern Communications* (Basic Books, 2004), 216.
54. Douglas M. O'Reagan, *Taking Nazi Technology: Allied Exploitation of German Science after the Second World War*, Kindle edition (Johns Hopkins University Press, 2019), 1458.
55. See Act of October 6, 1917, ch. 95, 40 Stat. 394 (1917 General Order no. 3 of February 1912 and General Order no. 64 of November 1917 (Relyea, 2008); Dorothy K.

McAllen, "National Security Policy Constraints on Technological Innovation: A Case Study of the Invention Secrecy Act of 1951," *Eastern Michigan University Master's Theses and Doctoral Dissertations*, paper 580 (2012), 53.

56. Rene D. Tegtmeyer, statement before Government Information and Individual Rights Subcommittee of the Committee on Government Operations, 1980, 6. http://www.fas.org/sgp/othergov/invention/private.pdf.

57. Tegtmeyer, statement before House subcommittee, 6.

58. Sabing H. Lee, "Protecting the Private Inventor under the Peacetime Provisions of the Invention Secrecy Act," *Berkeley Tech. LJ* 12 (1997): 356.

59. Lee, "Protecting the Private Inventor," 345.

60. Executive Order no. 8381 (March 22, 1940), https://fas.org/irp/offdocs/eo/eo-8381 .htm; McAllen, "National Security Policy Constraints on Technological Innovation," 53–54. Once the United States entered the war, it was extended for the duration of the conflict, like the World War I statute. Lee, "Protecting the Private Inventor," 350.

61. See H.R. Rep. No. 1540, 96th Cong., 2d Sess. 37 (1980).

62. See H.R. Rep. No. 1540, 96th Cong., 2d Sess. 37, 46–47 (1980).

63. McAllen, "National Security Policy Constraints on Technological Innovation," 54.

64. Invention Secrecy Act of 1951, 66 Stat. 3 (1952), codified as amended at 35 U.S.C. 181–88 (1994).

65. Invention Secrecy Act of 1951.

66. 35 U.S.C. § 181 para. 4 (1994).

67. President Truman declared a national emergency on December 16, 1950. Proclamation no. 2914, 3 C.F.R. 99 (1949–53). This national emergency was terminated by the National Emergencies Act of 1976, Pub. L. No. 94-412, Title I, 90 Stat. 1255 (1976).

68. Tegtmeyer, statement before House subcommittee, 36.

69. Tegtmeyer, statement before House subcommittee, 10. The decision as to whether there was sufficient need for secrecy was originally dispersed among a variety of executive departments. By the middle of the Cold War, this role was filled by the Armed Services Patent Advisory Board, which was composed of personnel at the defense agencies including the Army, Navy, Air Force, National Security Agency, Department of Energy, and NASA. In 1997 this role moved to a newly created agency in the Department of Defense, the Defense Threat Reduction Agency, and assigned to the Defense Technical Information Center. See also Eric B. Chen, "Technology Outpacing the Law: The Invention Secrecy Act of 1951 and the Outsourcing of U.S. Patent Application Drafting," 13 *Tex. Intell. Prop. L.J.* 351 (Spring 2005); G. W. Schulz, "Government Secrecy Orders on Patents Have Stifled More Than 5,000 Inventions," *Center for Investigative Reporting*, April 16, 2013, http:// www.wired.com/2013/04/gov-secrecy-orders-on-patents/. Patent applications were sent to that board if they fell within the classified list of sensitive technologies on the "Patent Security Category Review List."

70. Under Executive Order 9424, 3 C.F.R. 303 (1943–48), all government-owned or government-controlled interests in patent applications must be registered in the

Patent Office's Government Register. Therefore, by referring to the register, it is easy to determine whether a government property interest exists; 35 U.S.C. section 181.

71. McAllen, "National Security Policy Constraints on Technological Innovation," 25.

72. Tegtmeyer, statement before House subcommittee, 7.

73. Tegtmeyer, statement before House subcommittee, 43.

74. Tegtmeyer, statement before House subcommittee, 42–43.

75. Tegtmeyer, statement before House subcommittee, 43.

76. See also *Minnesota Mining and Manufacturing Co. v. Norton Co.*, 144 U.S.P.Q. (BNA) 272, 276 (N.D. Ohio 1965); *Blake v. Bassick Co.*, 146 U.S.P.Q. (BNA) 157–58 (N.D. Ill. 1963). See "Technology Outpacing the Law: The Invention Secrecy Act of 1951 and the Outsourcing of U.S. Patent Application Drafting," 13 *Tex. Intell. Prop. L.J.* 351 (Spring 2005). Judicial review is generally not available when the "agency action is committed to agency discretion by law"; see 5 U.S.C. § 701(a)(2) (1994).

77. Tegtmeyer, statement before House subcommittee, 6.

78. Tegtmeyer, statement before House subcommittee, 6.

79. See H.R. Rep. No. 1540, at 3. Tegtmeyer, statement before House subcommittee, 6.

80. The PTO assistant commissioner testified that "most applications under secrecy orders are related to government property interests." Tegtmeyer, statement before House subcommittee, 7, 38.

81. See S. Rep no. 1001, H.R. Rep. no. 1028, *supra* note 38, reprinted in 1952 U.S.C.C.A.N. at 1323; see also H.R. Rep. no. 1540, *supra* note 9.

82. See *Constant v. United States*, 617 F.2d 239, 244 (Ct. Cl. 1980); *Constant v. United States*, 1 Cl. Ct. 600, 609 (1982), aff'd, 714 F.2d 162 (Fed. Cir. 1983).

83. McAllen, "National Security Policy Constraints on Technological Innovation," 53.

84. McAllen, "National Security Policy Constraints on Technological Innovation," 60.

85. Steven Aftergood, "Invention Secrecy Increased in 2016," Federation of American Scientists, October 31, 2016. https://fas.org/blogs/secrecy/2016/10/invention-secrecy-2016.

86. In 2014 the inventors Budimir and Desanka Damnjanovic filed a lawsuit seeking compensation regarding a secrecy order that the US Air Force had imposed on them. Among their arguments was that the act was a violation of the First Amendment's right to free speech. However, the court did not reach the merits of the argument, dismissing the constitutional claim for procedural reasons. The plaintiffs ended up settling with the government for a lump sum of $63,000. Aftergood, at https://fas.org/blogs/secrecy/2016/10/invention-secrecy-2016.

87. Scott E. Chandler, "Rethinking Competition in Defense Acquisition," Lexington Institute, 2014; 2.

88. Interview with law firm partner, July 3, 2014.

89. Deutsch, "Consolidation," 139.

90. Rik Kirkland, "Don Rumsfeld Talks Guns and Butter," *Fortune*, November 18, 2002; Philip Ewing, "Ash Carter's Appeal to Silicon Valley: We're 'Cool' too," Politico, April 23, 2015. http://www.politico.com/story/2015/04/ash-carter-silicon-valley-appeal-117293.

91. Public Law 95–507. https://www.gpo.gov/fdsys/pkg/STATUTE-92/pdf/STATUTE-92-Pg1757.pdf.

92. Department of Defense Office of Small Business Programs, https://www.acq.osd.mil/osbp/osbp/about.html.

93. Ash Carter, "Rewiring the Pentagon: Charting a New Path on Innovation and Cybersecurity," Department of Defense, April 23, 2015, http://www.defense.gov/Speeches/Speech.aspx?SpeechID=1935.

94. Carter, "Rewiring the Pentagon."

95. Marcus Weisgerber, "Pentagon Sends an Engineer and a Navy SEAL to Woo Silicon Valley," Defense One, August 5, 2015. http://www.defenseone.com/management/2015/08/pentagon-engineer-and-navy-seal-woo-silicon-valley/118870/.

96. Allyson Versprille, "Defense Innovation Unit Tasked to Circumvent Traditional Acquisition System," *National Defense*, October 29, 2015. http://www.nationaldefensemagazine.org/blog/Lists/Posts/Post.aspx?ID=2001.

97. Dror Etzion and Gerald F. Davis, "Revolving Doors? A Network Analysis of Corporate Officers and US Government Officials," *Journal of Management Inquiry* 17, no. 3 (2008): 157–61.

98. Daniel Goure, "Incentivizing a New Defense Industrial Base," *Lexington Institute*, September 2015, 5–7.

99. Dombrowski and Gholz, *Buying Military Transformation*, 139.

100. Bryan. Bender, "From the Pentagon to the Private Sector," *Boston Globe*, December 26, 2010. http://www.boston.com/news/nation/washington/articles/2010/12/26/defense_firms_lure_retired_generals/?page=full.

101. Michael L. Slonecker, "Patently-O TidBits," Patently-O, March 27, 2007. http://www.-patentlyo.com/patent/2007/03/patentlyo_tidbi_2.html#comment-64529722.

102. Robert M. Chesney, "State Secrets and the Limits of National Security Litigation," *George Washington Law Review* 75 (2006): 1249.

103. Louis Fisher, "The State Secrets Privilege: Relying on Reynolds," *Political Science Quarterly* 122, no. 3 (2007): 392.

104. Fisher, "State Secrets," 408.

105. "Morning Edition: Administration Employing State Secrets Privilege at Quick Clip," National Public Radio, September 9, 2005. http://www.npr.org/templates/story/story.php?storyId=4838701. See also Chesney, "State Secrets," 1249, 1291–92.

106. 154 CONG. REc. S198 (daily ed. Jan. 23, 2008; statement of Sen. Kennedy.

107. Fisher, "State Secrets," 394.

108. *United States v. Reynolds*, 345 U.S. 1 (1953).

109. *United States v. Reynolds*, 345 U.S. 1 (1953).

110. See, for example, *Molerio v. FBI*, 749 F.2d 815, 822 (D.C. Cir. 1984); *Al-Haramain Islamic Found., Inc. v. Bush*, 507 F.3d 1190, 1203–04 (9th Cir. 2007).

111. *Kasza v. Browner*, 133 F.3d 1159, 1166 (9th Cir. 1998); *Clift v. U.S.*, 597 F.2d 826, 828–9 (2d Cir. 1979); *Crater*, 423 F.3d at 1267; Carrie Newton Lyons, "The State Secrets Privilege: Expanding Its Scope through Government Misuse," 11 Lewis & Clark L. Rev. 99 (2007); *White v. Raytheon*, 2008 U.S. Dist. LEXIS 102200 (D.Ma. 2008).

112. *General Dynamics Corporation v. U.S.*, 563 U.S. __ (2011) (consolidated with *Boeing v. U.S.*, no. 09–1302.

113. *Gulet Mohamed v. Holder*, case No. 1:11-CV-0050 (E.D.Va.), memorandum filed January 23, 2015.

114. 945 F.2d 1285, 1289 (4th Cir. 1991); *Clift v. United States*, 597 F.2d 826, 829 (2d Cir. 1979); *N.S.N. Int'l Indus. v. E.I. Dupont de Nemours & Co.*, 140 F.R.D. 275, 281 (S.D.N.Y. 1991); *Contra Halpern v. United States*, 258 F.2d 36, 44 (2d Cir. 1958); 808 F. Supp. 101, 102–03 (D. Conn. 1991).

115. Daniel R Cassman, "Keep It Secret, Keep It Safe: An Empirical Analysis of the State Secrets Doctrine," *Stanford Law Review* 67 (2015): 1199.

Chapter 4

1. Much of the material in this section first appeared in similar form in Robert Farley and Davida Isaacs, "The Stark Reality of Defense Contracting," *American Prospect*, May 10, 2010. http://prospect.org/article/stark-reality-defense-contract ing-0; Iron Man Wiki, "Crimson Dynamo," http://ironman.wikia.com/wiki/Cri mson_Dynamo.

2. Iron Man Wiki, "Crimson Dynamo."

3. *Iron Man 2*, directed by Jon Favreau, (2010; Buena Vista Home Entertainment, 2013), DVD.

4. Mike Lee, "Little-Known Sci-Fi Fact: Stan Lee Thought Marvel's Readers Would HATE Iron Man (at First)," *Blastr*, April 30, 2013, http://www.blastr.com/2013– 4-30/little-known-sci-fi-fact-stan-lee-thought-marvel%E2%80%99s-readers-would-hate-iron-man-first; Iron Man Wiki, "Crimson Dynamo."

5. Chris Morris, "10 Surprising Tetris Facts from the Game's Creator," CNBC, June 10, 2014. http://www.cnbc.com/2014/06/10/10-things-you-didnt-know.html.

6. Katherine C. Epstein, "Intellectual Property and National Security: The Case of the Hardcastle Superheater, 1905–1927," *History and Technology* 34, no. 2 (2018): 143.

7. Albert Guangzhou Hu and Gary H. Jefferson, "Returns to Research and Develop-ment in Chinese Industry: Evidence from State-Owned Enterprises in Beijing," *China Economic Review* 15, no. 1 (2004): 86–107.

8. Robert William Davies, Mark Harrison, and Stephen G. Wheatcroft, *The Eco-nomic Transformation of the Soviet Union, 1913–1945* (Cambridge University Press, 1994), 5.

9. Mark Harrison, "The Soviet Union: The Defeated Victor," in Mark Harrison, ed., *The Economics of World War II: Six Great Powers in International Comparison* (Cam-bridge University Press, 2000).

10. Andrew F Krepinevich and Barry D. Watts. *The Last Warrior: Andrew Marshall and the Shaping of Modern American Defense Strategy* (Basic Books, 2015), 150.

11. Alexander Hill, *The Red Army and the Second World War* (Cambridge University Press, 2016), 39.

12. Hill, *The Red Army*, 39.

13. Hill, *The Red Army*, 183–85.

14. "List of Most-Produced Aircraft," Wikipedia, May 15, 2018. https://en.wikipedia.org/wiki/List_of_most-produced_aircraft.

15. Jürgen Rohwer and Mikhail S. Monakov. *Stalin's Ocean-Going Fleet: Soviet Naval Strategy and Shipbuilding Programmes, 1935–1953* (Psychology Press, 2001), 185.

16. Hill, *The Red Army*, 39.

17. G. Scott Gorman, "The TU-4: The Travails of Technology Transfer by Imitation." *Air Power History* 45, no. 1 (1998): 16.

18. Harold J. Berman and John R. Garson. "United States Export Controls: Past, Present, and Future," *Columbia Law Review* 67, no. 5 (1967): 791–890; 792.

19. Paulina Galtsova, "Intellectual Property Reform in Russia: Analysis of Part Four of the Russian Civil Code" (master's thesis, Faculty of Law, Lund University, 2008), 12–13.

20. Galtsova, "Intellectual Property Reform," 13.

21. Galtsova, "Intellectual Property Reform," 19; Lisa D. Cook, "A Green Light for Red Patents? Evidence from Soviet Domestic and Foreign Inventive Activity, 1962 to 1991," thesis, Michigan State University (2011), 3–4.

22. John A. Martens, *Secret Patenting in the USSR and Russia* (Deep North Press, 2010), 90.

23. Martens, *Secret Patenting*, 145.

24. Martens, *Secret Patenting*, 148.

25. Martens, *Secret Patenting*, 150.

26. Stephen G. Brooks, *Producing Security: Multinational Corporations, Globalization, and the Changing Calculus of Conflict* (Princeton University Press, 2007), 126–27.

27. Katherine Amelia Siobhan Sibley, *Red Spies in America: Stolen Secrets and the Dawn of the Cold War* (University Press of Kansas, 2004).

28. Hill, *The Red Army*, 36.

29. Hill, *The Red Army*, 34.

30. Hill, *The Red Army*, 35.

31. Philip Hanson, "Soviet Industrial Espionage," *Bulletin of the Atomic Scientists* 43, no. 3 (1987): 25–29.

32. Hanson, "Soviet Industrial Espionage," 25–29.

33. Hanson, "Soviet Industrial Espionage," 25–29.

34. Hanson, "Soviet Industrial Espionage," 28.

35. Ian Anthony, "Economic Dimensions of Soviet and Russian Arms Exports," in Ian Anthony, ed., *Russia and the Arms Trade* (Oxford University Press on Demand, 1998), 74; Kirshin Yuriy, "Conventional Arms Transfers during the Soviet Period," in Anthony, *Russia and the Arms Trade*, 48.

36. Robert M. Cutler, Laure Després, and Aaron Karp, "The Political Economy of East–South Military Transfers," *International Studies Quarterly* 31, no. 3 (September 1987): 273–99; 278.

37. Cutler et al., "Political Economy," 275.

38. Cutler et al., "Political Economy," 275.

39. Jennifer Anderson, "The Limits of Sino-Russian Partnership." *Adelphi Paper* 315 (1997).

40. Santosh Mehrotra, *India and the Soviet Union: Trade and Technology Transfer* (Cambridge University Press, 1990), 22.

41. Charlotte Mathieu, "Assessing Russia's Space Cooperation with China and India: Opportunities and Challenges for Europe." *Acta Astronautica* 66, no. 3–4 (2010): 355–61.

42. Cutler et al., "Political Economy," 285.

43. Cutler et al., "Political Economy," 291.

44. Vladimir Karznorov, "Algeria Returns 'Faulty' MiG-29s," *Flight Global*, February 25, 2008, https://www.flightglobal.com/news/articles/algeria-returns-faulty-mig -29s-221771/.

45. Louis-Marie Clouet, "Rosoboronexport, Spearhead of the Russian Arms Industry." *Russie Nei Visions* 22 (2007): 4.

46. Niclas Rolander, "Russia's Arms Exports Grow," *Wall Street Journal*, March 16, 2014, https://www.wsj.com/articles/SB10001424052702303287804579443102 858150332.

47. P. R Chari, "Indo-Soviet Military Cooperation: A Review," *Asian Survey* 19, no. 3 (1979): 230–44; 232.

48. Brooks, *Producing Security*, 95.

49. Linda Jakobson, Paul Holtom, Dean Knox, and Jingchao Peng, *China's Energy and Security Relations with Russia: Hopes, Frustrations, Uncertainties* (SIPRI, October 2011), 14.

50. Jakobson et al., *China's Energy and Security*.

51. Jakobson et al., *China's Energy and Security*.

52. Jyotsna Bakshi, "India-Russia Defence Cooperation." *Strategic Analysis* 30, no. 2 (2006): 450–51.

53. "Aircraft and Airspace Industry of Ukraine (2007)," Ministry of Economic Development and Trade, http://www.ukrexport.gov.ua/eng/economy/ukr/203.html.

54. Jess McHugh, "France and Russia Mistral Deal: $1.3 Billion Settlement Reached Over Aircraft Carriers," *International Business Times*, July 10, 2015, http://www .ibtimes.com/france-russia-mistral-deal-13-billion-settlement-reached-over -aircraft-carriers-2003044.

55. Franz-Stefan Gady, "Will India, Russia Co-Develop a New 5th Generation Stealth Fighter?" *The Diplomat*, March 25, 2017, http://thediplomat.com/2017/03/will -india-russia-co-develop-a-new-5th-stealth-fighter/.

56. Nicholas Gvosdev, "Why America Should Really Fear Russia's Armata T-14 Tank," *National Interest*, May 8, 2015, http://nationalinterest.org/feature/why-america -should-really-fear-russias-armata-t-14-tank-12836?page=2.

57. "Unlimited Upgrades & Value Mean Russian T-14 Armata Tank Is SExport Gem: Developer," RT, May 26, 2015, http://rt.com/news/262105-armata-t14-upgrades -export/.

58. Mark Episkopos, "Russia's New Armata Tank: The Best Tank in the World?" *National Interest*, April 23, 2019, https://nationalinterest.org/blog/buzz/russias-new -armata-tank-best-tank-world-53837.

59. Rakesh Krishnan Simha, "Rise of the Clones: Chinese Knockoffs Undercut Russian Arms Exports," *Russia beyond the Headlines*, August 11, 2015, http://rbth.com /blogs/2015/08/11/rise_of_the_clones_chinese_knockoffs_undercut_russian _arms_exports_48345.html.

60. Galtsova, *Intellectual Property Reform*, 6.
61. Galtsova, *Intellectual Property Reform*, 6.
62. Galtsova, *Intellectual Property Reform*, 6.
63. Sergey Budylin and Yulia Osipova, "Total Upgrade: Intellectual Property Law Reform in Russia," *Columbia Journal of East European Law* 1, no. 1 (2007): 8.
64. Budylin and Osipova. "Total Upgrade," 8.
65. Budylin and Osipova, "Total Upgrade," 9.
66. Sergey Kortunov, "The Influence of External Factors on Russia's Arms Export Policy," in Ian Anthony, ed., *Russia and the Arms Trade* (Oxford University Press, 1998), 105.
67. Jakobson et al., *China's Energy and Security Relations with Russia*, 21.
68. Rakesh Krishnan Simha, "F-35B: Born in the USSR," *Russia and India Report*, June 10, 2013, http://in.rbth.com/blogs/2013/06/07/f-35b_born_in_the_ussr_25935.
69. Zaal Tchkuaseli, "Yakolev Yak-141," http://www.military-today.com/aircraft/yak_141.htm.
70. Simha, "F-35B."
71. Jim Smith, "Before F-35: My Role in the Advanced Short Take-Off/Vertical Landing (ASTOVL) Project," Hush Kit, May 11, 2019, https://hushkit.net/2019/05/11/before-f-35-my-role-in-the-advanced-short-take-off-vertical-landing-astovl-project/.
72. Eugene Gerden, "Russian Government Designs New Strategy in Field Of IP," IP-Watch.com, July 13, 2015, http://www.ip-watch.org/2015/07/13/russian-government-designs-new-strategy-in-field-of-ip/.
73. Gerden, "Russian Government Designs New Strategy."
74. Tai Ming Cheung, *Fortifying China: The Struggle to Build a Modern Defense Economy*, Kindle edition (Cornell University Press, 2013), 721.
75. Cheung, *Fortifying China*, 992.
76. Cheung, *Fortifying China*, 782.
77. Cheung, *Fortifying China*, 908.
78. John R. Allison and Lianlian Lin, "Evolution of Chinese attitudes toward Property Rights in Invention and Discovery," *University of Pennsylvania Journal of International Economic Law.* 20 (1999): 742; William P. Alford, *To Steal a Book Is an Elegant Offense: Intellectual Property Law in Chinese Civilization.* (Stanford University Press, 1995).
79. Richard S. Horowitz, "Beyond the Marble Boat: The Transformation of the Chinese Military, 1850–1911," in David A. Graff and Robin Higham, *A Military History of China* (Westview Press, 2002), 153–74.
80. Allison and Lin, "Evolution of Chinese Attitudes," 747.
81. Allison and Lin, "Evolution of Chinese Attitudes," 750.
82. Cheung, *Fortifying China*, 986.
83. Cheung, *Fortifying China*, 986.
84. Tai Ming Cheung, Thomas G. Mahnken, and Andrew L. Ross, "Frameworks for Analyzing Chinese Defense and Military Innovation," in Tai Ming Cheung, ed., *Forging China's Military Might: A New Framework for Assessing Innovation*, Kindle edition (Johns Hopkins University Press, 2014), 666–73.

85. For fuller discussion, see William C Hannas. James Mulvenon, and Anna B. Puglisi, *Chinese Industrial Espionage: Technology Acquisition and Military Modernization* (Routledge, 2013); Keith Crane, Roger Cliff, Evan Medeiros, James Mulvenon, and William Overholt, *Modernizing China's Military: Opportunities and Constraints* (Rand, 2005).

86. Carlo Kopp, "Chengdu J-10: Technical Report APA-TR-2007–0701," *Air Power Australia*, January 27, 2014. http://www.ausairpower.net/APA-Sinocanard.html.

87. Hannas et al., *Chinese Industrial Espionage*.

88. Tai Ming Cheung, "The Chinese Defense Economy's Long March from Imitation to Innovation," in *China's Emergence as a Defense Technological Power* (Routledge, 2013), 42.

89. Wendell Minnick, "Russia Admits China Illegally Copied Its Fighter," *Defense News*, February 13, 2009, http://minnickarticles.blogspot.com/2009/10/russia-admits-china-illegally-copied.html. See also chapter 5 in this book.

90. Cheung, "Long March," 36.

91. Stuart Macdonald, "Nothing Either Good or Bad: Industrial Espionage and Technology Transfer," *International Journal of Technology Management* 8, no. 1/2 (1993).

92. Wendell Minnick, "Experts: China Still Lags West in Advanced Aircraft Technologies," *Defense News*, August 3, 2014, http://archive.defensenews.com/article/20140803/DEFREG03/308030011/Experts-China-Still-Lags-West-Advanced-Aircraft-Technologies.

93. Karl W. Eikenberry, *Explaining and Influencing Chinese Arms Transfers* (DIANE Publishing, 1995), 9.

94. Steve Davies, *Red Eagles: America's Secret MiGs* (Osprey, 2008).

95. "From MiG-21 to J-7," Globalsecurity.org, November 11, 2007, http://www.globalsecurity.org/military/world/china/j-7-dev.htm.

96. Davies, *Red Eagles*.

97. SIPRI, "Arms Transfer Database," https://www.sipri.org/databases/armstransfers.

98. Allison and Lin, "Evolution of Chinese Attitudes," 755.

99. Cheung, *Fortifying China*, 1421.

100. Cheung, *Fortifying China*, 390.

101. Samm Tyroler-Cooper, and Alison Peet, "The Chinese Aviation Industry: Techno-Hybrid Patterns of Development in the C919 Program," in Tai Ming Cheung, ed., *China's Emergence as a Defense Technological Power*, Kindle edition (Routledge, 2013), 103.

102. Tyroler-Cooper and Peet, "The Chinese Aviation Industry," 103.

103. Cheung, *Fortifying China*, 1996.

104. Cheung, *Fortifying China*, 1496, 2000, 2130.

105. Cheung, *Fortifying China*, 2147.

106. Daniel Alderman, Lisa Crawford, Brian Lafferty, and Aaron Shraberg, "The Rise of Chinese Civil-Military Integration," in Tai Ming Cheung, ed., *Forging China's Military Might: A New Framework for Assessing Innovation*, 2646.

107. Alderman et al., "The Rise of Chinese Civil-Military Integration," 2728.

108. Alderman et al., "The Rise of Chinese Civil-Military Integration," 2344.

109. Alderman et al., "The Rise of Chinese Civil-Military Integration," 2332.

110. Cheung, "Long March," 42.

111. "JF-17 Thunder- Foreign Sales," GlobalSecurity.org, http://www.globalsecurity
.org/military/world/pakistan/jf-17-sales.htm.

112. James K. Sebenius and Rebecca Hulse, "Charlene Barshefsky," HBS 9-801-421
(*Harvard Business Review*, 2013), 7–8; Matthew Dresden, "China Copyrights: No,
You Can't Call It Fair Use," China Law Blog, February 28, 2017, http://www.china
lawblog.com/2017/02/china-copyrights-and-fair-use.html.

113. Tom Gjelten, "U.S. Turns Up Heat On Costly Commercial Cybertheft In China,"
National Public Radio, May 7, 2013, http://www.npr.org/2013/05/07/181668369
/u-s-turns-up-heat-on-costly-commercial-cyber-theft-in-china.

114. Keith Bradsher, "Trump Administration Is Said to Open Broad Inquiry into China's
Trade Practices," *New York Times*, August 1, 2017, https://www.nytimes.com/2017
/08/01/business/trump-china-trade-intellectual-property-section-301.html.

115. Peter K. Yu, "From Pirates to Partners: Protecting Intellectual Property in China
in the Twenty-First Century," *American University Law Review.* 50 (2001): 131.

116. State Intellectual Property Office of the People's Republic of China, "White Paper
on the Intellectual Property Rights Protection in China in 1998," http://english
.sipo.gov.cn/laws/whitepapers/200804/t20080416_380349.html.

117. William Weightman, "China's Progress on Intellectual Property Rights (Yes
Really)," *The Diplomat*, January 20, 2018, https://thediplomat.com/2018/01/chinas
-progress-on-intellectual-property-rights-yes-really/.

118. Weightman, "China's Progress."

119. Alderman et al., "The Rise of Chinese Civil-Military Integration," 2546.

120. Alderman et al., "The Rise of Chinese Civil-Military Integration," 2563.

121. Alderman et al., "The Rise of Chinese Civil-Military Integration," 2563.

122. Alderman et al., "The Rise of Chinese Civil-Military Integration," 2563.

123. Alderman et al., "The Rise of Chinese Civil-Military Integration," 2469.

124. Alderman et al., "The Rise of Chinese Civil-Military Integration," 2454.

125. Alderman et al., "The Rise of Chinese Civil-Military Integration," 2465.

126. Joe Katzman, "The Hyundaization of the Global Arms Industry," *Wall Street Journal*, April 5, 2015, http://www.wsj.com/articles/joe-katzman-the-hyundaization
-of-the-global-arms-industry-1428271215.

127. Richard A. Bitzinger, "The Global Arms Trade and the Hyundaization Threat," *The
Diplomat*, April 15, 2015, http://thediplomat.com/2015/04/global-arms-trade-and
-the-hyundaization-threat/.

128. "K1A1 Main Battle Tank, South Korea," http://www.army-technology.com/proj
ects/k1/.

129. "XK-2 Heukpyo [Black Panther] Main Battle Tank," Globalsecurity.org, November
24, 2014, http://www.globalsecurity.org/military/world/rok/xk-2.htm.

130. "Intellectual Property Rights in Korea," Intellectual Property Office, 2013,
https://www.gov.uk/government/uploads/system/uploads/attachment_data/file
/456362/IP_rights_in_Korea.pdf.

131. Ching-in Moon, "South Korea and International Compliance Behavior: The WTO

and IMF in Comparative Perspective," in Edward C. Luck and Michael W. Doyle, eds., *International Law and Organization: Closing the Compliance Gap* (Rowman and Littlefield, 2004), 69.

132. Office of the United States Trade Representative, "Intellectual Property Rights in U.S.-South Korea Trade Agreement," https://ustr.gov/uskoreaFTA/IPR.

133. "Intellectual Property Rights in Korea."

134. Moon, "South Korea and International Compliance Behavior," 71.

135. Lee, "South Korea's Export Control System," Stockholm International Peace Research Institute, November 2013, http://books.sipri.org/files/misc/SIPRIBP1311.pdf.

136. Lee, "South Korea's Export Control System," 3.

137. Zahra Ullah, "How Samsung Dominates South Korea's Economy," CNNtech, February 17, 2017, http://money.cnn.com/2017/02/17/technology/samsung-south-korea-daily-life/.

138. *Apple Inc. v. Samsung Electronics Co. Ltd. et al.* United States District Court, Northern District of California. August 11, 2012, http://cand.uscourts.gov/lhk/applev samsung.

139. Michael A. Cusumano, "The Apple-Samsung Lawsuits," *Communications of the ACM* 56, no. 1 (2013): 28–31.

140. Office of the United States Trade Representative, "United States-Korea Free Trade Agreement: Final Text," https://ustr.gov/trade-agreements/free-trade-agreements/korus-fta/final-text.

141. James Edwards, "Calling Out South Korean IP Theft on Fifth Anniversary of Trade Deal," *Washington Examiner*, April 14, 2017, https://www.washingtonexaminer.com/calling-out-south-korean-ip-theft-on-fifth-anniversary-of-trade-deal.

142. "South Korea: Defense Market Overview," Pennsylvania's Department of Community and Economic Development, Office of International Business Development, 2015, http://community.newpa.com/download/business/trade/market_research/South_Korea_-_Pennsylvania_Defense_Market_Overview_2015.pdf 11–14.

143. "Korea Aerospace Industries LTD," 4-Traders, http://www.4-traders.com/KOREA-AEROSPACE-INDUSTRIE-10202838/company/.

144. "SIPRI Arms Transfers Database." Stockholm International Peace Research Institute, December 23, 2015. http://armstrade.sipri.org/armstrade/page/trade_register.php.

145. Jaewon Lee, "South Korea's Export Control System."

146. Maharani Curie and C. S. L Koh, "Bracing for Impact: Fifth-Generation Jet Fighter Programmes in Asia," RSIS Commentaries, no. 045 (2010). https://dr.ntu.edu.sg/bitstream/handle/10220/6546/RSIS0452010.pdf?sequence=1.

147. Choe Sang-Hun, "South Korea Replaces Top Security Adviser over Fighter Jet Program," *New York Times*, October 19, 2015. http://www.nytimes.com/2015/10/20/world/asia/south-korea-replaces-top-security-adviser-over-fighter-jet-program.html?smid=tw-nytimesworld&smtyp=cur&_r=0.

148. "U.S. Refuses Key Technology Transfers for Korean KF-X Fighter Program," Aviation Analysis Wing, September 29, 2015. http://www.aviationanalysis.net/2015/09/us-refuses-key-technology-transfer-for-kfx-fighter-program.html.

149. "U.S. Refuses Key Technology Transfers."

150. Dave Majumdar, "Will America Help South Korea Build Lethal Stealth Fighters?" *National Interest*, October 14, 2015, http://www.nationalinterest.org/blog/the-bu zz/will-america-help-south-korea-build-lethal-stealth-fighters-14078.

151. Kang Seung-Woo, "US Strikes Down T-50 Exports to Uzbekistan," *Korea Times*, October 26, 2015, http://www.koreatimes.co.kr/www/news/nation/2015/10/116 _189398.html.

152. Kang, "US Strikes Down T-50 Exports to Uzbekistan."

153. "Turkey Selects K2 Black Panther Main Battle Tank for its Altay Program," Deagel .com, July 31, 2008, http://www.deagel.com/news/Turkey-Selects-K2-Black-Pan ther-Main-Battle-Tank-for-its-Altay-Program_n000006970.aspx.

154. Kelvin Wong, "DX Korea 2016: Hyundai Rotem Readies K806 and K808 Wheeled Armoured Vehicle Production for South Korean Army," *IHS Jane's 360*, September 9, 2016, https://web.archive.org/web/20160910130030/http://www.janes.com /article/63586/dx-korea-2016-hyundai-rotem-readies-k806-and-k808-wheeled -armoured-vehicle-production-for-south-korean-army.

155. Gordon Lubold, "Is South Korea Stealing U.S. Military Secrets? Their Tanks, Missiles, and Electronic Warfare Gear Look an Awful Lot Like Ours," *Foreign Policy*, October 28, 2013, https://foreignpolicy.com/2013/10/28/is-south-korea-stealing -u-s-military-secrets/.

156. Lubold, "Is North Korea Stealing U.S. Military Secrets?"

157. "Chang Bogo Class," Military-Today.com, http://www.military-today.com/navy /chang_bogo_class.htm; Oh Young-Jin, "Korea Wins $1 Billion Indonesia Deal," *Korea Times*, December 20, 2011, http://www.koreatimes.co.kr/www/news/biz/20 11/12/123_101248.html.

Chapter 5

1. Carlo Kopp, "Fulcrum and Flanker: The New Look in Soviet Air Superiority," *Australian Aviation*, May/June 1990, http://www.ausairpower.net/Profile-Fulcrum-Fl anker.html.

2. Kopp, "Fulcrum and Flanker."

3. Richard Rousseau, "The Torturous Sino-Russian Arms Trade-Analysis," *Eurasia Review*, June 9, 2012, http://www.eurasiareview.com/09062012-the-tortuous-si no-russian-arms-trade-analysis/.

4. "J-11," *GlobalSecurity.org*, February 5, 2015, http://www.globalsecurity.org/military /world/china/j-11.htm.

5. Linda Jakobsen, Paul Holtom, Dean Knox, and Jingchao Peng, *China's Energy and Security Relations with Russia: Hopes, Frustrations, Uncertainties*, SIPRI, October 2011, 14.

6. Wendell Minnick, "Russia Admits China Illegally Copied Its Fighter," *Defense News*, February 13, 2009, http://minnickarticles.blogspot.com/2009/10/russia-ad mits-china-illegally-copied.html.

7. Minnick, "Russia Admits China Illegally Copied Its Fighter."

8. Maria Chapligina, "Russia Downplays Chinese J-15 fighter Capabilities," *RIA*

Novosti, June 4, 2010, http://en.ria.ru/military_news/20100604/159306694 .html.

9. John Pomfret, "Military Strength Is Eluding China," *Washington Post*, December 25, 2010, http://www.washingtonpost.com/wp-dyn/content/article/2010/12/24 /AR2010122403009.html.

10. Jeffery Lin and Peter W. Singer. "The J-11D Surprise: China Upgrades Russian Flanker Fighters on Its Own," *Eastern Arsenal: Popular Science*, May 4, 2015, http:// www.popsci.com/j-11d-surprise-china-upgrades-russian-flanker-fighters-its -own.

11. Lin and Singer, "The J-11D Surprise."

12. Gabe Collins and Andrew Erickson, "Jet Engine Development in China: Indige-nous High-Performance Turbofans Are a Final Step toward Fully Independent Fighter Production," *China SignPost*, June 26, 2011, http://www.chinasignpost.com /2011/06/26/jet-engine-development-in-china-indigenous-high-performance -turbofans-are-a-final-step-toward-fully-independent-fighter-production/.

13. Wendell Minnick, "Russia: No Deal on Sale of Fighters, Subs to China," *Defense News*, March 25, 2013, http://www.defensenews.com/article/20130325/DEFREG 03/303250014/.

14. Franz Stefan-Gady, "China to Receive 10 Su-35 Advanced Fighter Jets in 2017," *The Diplomat*, February 8, 2017, http://thediplomat.com/2017/02/china-to-receive -10-su-35-advanced-fighter-jets-in-2017/.

15. Rousseau, "Sino-Russian Arms."

16. "Russian Tank Maker Doesn't Want Armata Sold to Beijing," *Want China Times*, January 25, 2015, http://www.wantchinatimes.com/news-subclass-cnt.aspx?id= 20150125000076&cid=1101.

17. Franz-Stefan Gady, "Russia Will Not Mass-Produce T-14 Armata Main Battle Tank," *The Diplomat*, August 1, 2018, https://thediplomat.com/2018/08/russia-will -not-mass-produce-t-14-armata-main-battle-tank/.

18. Rachel Stohl and Suzette Grillot. *International Arms Trade* (Wiley, 2013), 13.

19. David C. Evans, and Mark R. Peattie. *Kaigun: Strategy, Tactics, and Technology in the Imperial Japanese Navy 1887–1941* (Seaforth, 2012).

20. Stohl and Grillot, *International Arms Trade*, 18.

21. Stephanie G. Neuman, "Power, Influence, and Hierarchy: Defense Industries in a Unipolar World," in Richard A. Bitzinger, ed., *The Modern Defense Industry: Politi-cal, Economic, and Technological Issues: Political, Economic, and Technological Issues* (ABC-CLIO, 2009) 82, 63.

22. Neuman, "Power, Influence, and Hierarchy," 62.

23. See, for example, Sushant Singh, "What the Rafale Fighter Jet Deal Means for India," *Indian Express*, April 14, 2015, http://indianexpress.com/article/india/rafale -fighter-jet-deal-what-does-it-mean-for-india-and-what-is-the-way-forward/.

24. Stohl and Grillot, *International Arms Trade*, 88.

25. T. V. Paul, "Influence through Arms Transfers: Lessons from the US-Pakistani Re-lationship." *Asian Survey* 32, no. 12 (1992): 1078–92.

26. Sahni Varun, "Not Quite British: A Study of External Influences on the Argentine Navy," *Journal of Latin American Studies* 25, no. 3 (1993): 489–513; 507.

27. Katherine C. Epstein, *Torpedo* (Harvard University Press, 2014), 15.

28. Gareth Jennings, "Saab Not Selling Gripens to Argentina, Official Says," *IHS Jane's 360*, April 28, 2015, http://www.janes.com/article/51023/saab-not-selling-gripens -to-argentina-official-says.

29. "Report: U.S. Downgrades Saudi Arms Deal over Israeli Concerns," *Haaretz*, August 9, 2010, http://www.haaretz.com/news/diplomacy-defense/report-u-s-down grades-saudi-arms-deal-over-israeli-concerns-1.306886.

30. "Council of Ministers Declaration on China," June 26–27, 1989. http://www.sipri .org/databases/embargoes/eu_arms_embargoes/china/eu-council-of-minister -declaration-on-china.

31. Steve Davies, *Red Eagles: America's Secret MiGs* (Osprey, 2012).

32. Arms Control Association, "International Code of Conduct against Ballistic Missile Proliferation," https://www.armscontrol.org/documents/icoc.

33. Department of Commerce, Bureau of Industry and Security, "Chemical Weapons Convention: Export and Import Requirements," *Chemical Weapons Convention Bulletin*, May 2004, http://www.cwc.gov/outreach_industry_publications_cwc007 .html.

34. United Nations, "United Nations Office for Disarmament Affairs," http://www.un .org/disarmament/convarms/salw/.

35. Eric Sofge, "A Ban on Autonomous Killer Robots is Inevitable," *Popular Science*, August 5, 2015, http://www.popsci.com/ban-autonomous-killer-robots-inevitable.

36. Catherine Putz, "US Stops Export of Korean Fighter Jets to Uzbekistan," *The Diplomat*, October 27, 2015, http://thediplomat.com/2015/10/us-stops-export-of -korean-fighter-jets-to-uzbekistan/.

37. Robert M. Cutler, Laure Després, and Aaron Karp, "The Political Economy of East–South Military Transfers," *International Studies Quarterly* 31, no. 3 (September 1987): 273–99, accessed December 23, 2015, at http://www.robertcutler.org /download/html/ar87isq.html.

38. Robert Farley, "Making the PLAN: A Concise History of the People's Liberation Army Navy," in Donovan C. Chau and Thomas Kane, eds., *China and International Security* (ABC Clio, 2014), 99–118.

39. For a review of this literature, see Jian-Ye Wang and Magnus Blomström. "Foreign Investment and Technology Transfer: A Simple Model," *European Economic Review* 36, no. 1 (1992): 137–55; David J. Teece, "Technology Transfer by Multinational Firms: The Resource Cost of Transferring Technological Know-How," *Economic Journal* 87, no. 346 (1977): 242–61; Amy Jocelyn Glass, Amy Jocelyn, and Kamal Saggi, "International Technology Transfer and the Technology Gap," *Journal of Development Economics* 55, no. 2 (1998): 369–98; Jeremy Howells, "Tacit Knowledge, Innovation and Technology Transfer," *Technology Analysis & Strategic Management* 8, no. 2 (1996): 91–106.

40. Phillip Finnegan, "The Evolution of Defense Hierarchies," in Richard A. Bitzinger, ed., *The Modern Defense Industry: Political, Economic, and Technological Issues* (ABC Clio, 2009), 95–106; Douglas M. O'Reagan, *Taking Nazi Technology: Allied Exploitation of German Science after the Second World War*, Kindle edition (Johns Hopkins University Press, 2019), 1097–1100.

41. Stephanie G. Neuman, "Power, Influence, and Hierarchy: Defense Industries in a Unipolar World," in Bitzinger, *The Modern Defense Industry*, 82.

42. Ankit Panda, "India Will Buy 36 Ready-to-Fly Dassault Rafale Fighters from France," *The Diplomat*, April 13, 2015, http://thediplomat.com/2015/04/india-will -buy-36-ready-to-fly-dassault-rafale-fighters-from-france/.

43. Guillaume Lecompte-Boinet, "Dassault Agrees Rafale Workshare with India," *AIN Online Defense*, March 13, 2014, http://www.ainonline.com/aviation-news /defense/2014–03–13/dassault-agrees-rafale-workshare-india.

44. Caroline Bruneau, "French MoD Backs Dassault's Position on Indian Rafale," *Aviation Week*, February 9, 2015, http://aviationweek.com/blog/french-mod-backs-da ssaults-position-indian-rafale.

45. James Hasik, "Better Buying Power or Better Off Not? Purchasing Technical Data for Weapon Systems," *Defense Acquisition Research Journal* 21, no. 3 (July 2014): 697.

46. Ankit Panda, "A First: Japan Will Share Classified Submarine Technical Data with Australia," *The Diplomat*, May 7, 2015, http://thediplomat.com/2015/05/a-first-ja pan-will-share-classified-submarine-technical-data-with-australia/.

47. Franz-Stefan Gady, "Australia, France Sign Framework Agreement for $35.5 Billion Submarine Deal," *Diplomat: APAC*, February 26, 2019, https://thediplomat .com/2019/02/australia-france-sign-framework-agreement-for-35–5-billion -submarine-deal/.

48. Stephen G. Brooks, *Producing Security: Multinational Corporations, Globalization, and the Changing Calculus of Conflict* (Princeton University Press, 2007), 65.

49. Brooks, *Producing Security*, 65.

50. Paul J. Dunne, "Developments in the Global Arms Industry from the End of the Cold War to the mid-2000s," in *The Modern Defense Industry*, 19–21.

51. Theodore H. Moran, "The Globalization of America's Defense Industries: Managing the Threat of Foreign Dependence," *International Security* 15, no. 1 (1990): 64.

52. Rachel Stohl and Suzette Grillot, *International Arms Trade* (Wiley, 2013), 37.

53. Stohl and Grillot, *International Arms Trade*, 16.

54. Neuman, "Power, Influence, and Hierarchy," 66.

55. Keith Hayward, "The Globalization of Defense Industries," in *The Modern Defense Industry*, 108.

56. Ron Matthews and Curie Maharani, "The Defense Iron Triangle Revisited," in *The Modern Defense Industry*, 51, 109.

57. Davida H. Isaacs, "Not All Property Is Created Equal: Why Modern Courts Resist Applying the Takings Clause to Patents, and Why They Are Right to Do So," *George Mason Law Review*. 15 (2007): 1.

58. Ankit Panda, "India's Defense FDI Limit Will Hinder Technology Transfer: US Official," *The Diplomat*, September 21, 2015, http://thediplomat.com/2015/09/in dias-defense-fdi-limit-will-hinder-technology-transfer-us-official/.

59. Ankit Panda, "India's Defense FDI Limit."

60. Biswajit Dahr and T. C. James, "USTR's Investigations on IP Rights against India: Is There a Tenable Case?" *Intellectual Property Watch*, October 20, 2015, http://www.ip

-watch.org/2014/10/20/ustrs-investigations-on-ip-rights-against-india-is-there
-a-tenable-case/.

61. William New, "India, US Take Stock of Work on IP: To Boost Copyright, Trade
    Secrets," *Intellectual Property Watch*, October 30, 2015, http://www.ip-watch.org
    /2015/10/30/india-us-take-stock-of-joint-work-on-ip-bilateral-copyright-to-get
    -boost/.

62. Brooks, *Producing Security*, 113.

63. Sydney J. Freedberg Jr., "Navy Warship Is Taking 3D Printer to Sea; Don't Expect
    a Revolution," *Breaking Defense*, April 22, 2014, http://breakingdefense.com/2014
    /04/navy-carrier-is-taking-3d-printer-to-sea-dont-expect-a-revolution/.

64. Sarah Anderson, "China's PLA Navy Deploys 3D Printers onboard Warships to Re-
    place Small Parts," *3DPrint.com*, January 8, 2015, http://3dprint.com/35981/china
    -pla-navy-3d-printing/.

65. S. Bradshaw, A. Bowyer, and P. Haufe, "The Intellectual Property Implications of
    Low-Cost 3D Printing," *SCRIPTed* 7, no. 1 (2010), http://www.law.ed.ac.uk/ahrc
    /script-ed/vol7–1/bradshaw.asp.

66. Justin E. Pierce and Steven J. Schwarz. "IP Strategies and the Rise of 3D Printing,"
    Venable LLP, April 14, 2015, https://www.venable.com/ip-strategies-for-the-rise
    -of-3d-printing-04–14–2015/?utm_source=Mondaq&utm_medium=syndica
    tion&utm_campaign=View-Original.

67. "What Is 3D Printing?" *3DPrinting.com*, 2015, http://3dprinting.com/what-is-3d
    -printing/.

68. "What Is 3D Printing?"

69. Bradshaw, "Low-Cost 3D Printing."

70. Pierce and Schwarz, "IP Strategies."

71. Pierce and Schwarz, "IP Strategies."

72. Pierce and Schwarz, "IP Strategies."

73. Pierce and Schwarz, "IP Strategies."

74. Erin Carson, "3D Printing: Overcoming the Legal and Intellectual Property
    Issues," ZDNet, August 1, 2014, http://www.zdnet.com/article/3d-printing-over
    coming-the-legal-and-intellectual-property-issues/.

75. Carson, "3D Printing."

76. T. E. Halterman, "Zortrax 3D Printers Used to Modernize Iconic MiG-29 Fighter
    Jets," 3DPrint.com, August 10, 2015, https://3dprint.com/88176/zortrax-3d-pri
    nters-used-to-modernize-iconic-mig-29-fighter-planes/

77. Francis Cervano, "Export Controls and Their Relationship to National Defense
    Industries," in Bitzinger, *The Modern Defense Industry*, 244.

78. Charles A. Lofgren, "United States v. Curtiss-Wright Export Corporation: An His-
    torical Reassessment." *Yale Law Journal* 83, no. 1 (1973): 1–32.

79. Dorothy K. McAllen, "National Security Policy Constraints on Technological
    Innovation: A Case Study of the Invention Secrecy Act of 1951," *Master's Theses
    and Doctoral Dissertations*, paper 580 (2012): 53–54. http://commons.emich.edu
    /cgi/viewcontent.cgi?article=1956&context=theses.

80. Cervano, "Export Controls," 245.

81. Wende A. Wrubel, "Toshiba-Kongsberg Incident: Shortcomings of Cocom, and

Recommendations for Increased Effectiveness of Export Controls to the East Bloc," *American University International Law Review* 4 (1989): 245–46.

82. Wrubel, "Toshiba-Kongsberg," 245.

83. Mario Daniels, "Restricting the Transnational Movement of 'Knowledgeable Bodies': The Interplay of US Visa Restrictions and Export Controls in the Cold War," in John Krige, ed., *How Knowledge Moves: Writing the Transnational History of Science and Technology* (University of Chicago Press, 2019), 35.

84. Daniels, "Knowledgeable Bodies," 42.

85. Daniels, "Knowledgeable Bodies," 43.

86. Daniels, "Knowledgeable Bodies," 48.

87. David Zweig, Chen Changgui, and Stanley Rosen, "Globalization and Transnational Human Capital: Overseas and Returnee Scholars to China," *China Quarterly* 179 (2004): 735–57.

88. John Krige, "Export Controls as Instruments to Regulate Knowledge Acquisition in a Globalizing Economy," in John Krige, ed., *How Knowledge Moves: Writing the Transnational History of Science and Technology* (University of Chicago Press, 2019), 65.

89. Krige, "Export Controls," 69.

90. Krige, "Export Controls," 80.

91. Cervano, "Export Controls," 245.

92. Cervano, "Export Controls," 245.

93. Cervano, "Export Controls," 245.

94. Cora Currier, "In Big Win for Defense Industry, Obama Rolls Back Limits on Arms Exports," ProPublica, October 14, 2013, https://www.propublica.org/article/in-big -win-for-defense-industry-obama-rolls-back-limits-on-arms-export.

95. Jian Bin Gao and David Hardin, "The Export Control Risks of US-China Technology Collaboration," *China Business Review*, October 1, 2012, http://www.china businessreview.com/the-export-control-risks-of-us-china-technology-colla boration/.

96. Stephen D. Kelly, "Curbing Illegal Transfers of Foreign-Developed Critical High Technology from CoCom Nations to the Soviet Union: An Analysis of the Toshiba-Kongsberg Incident." *Boston College International and Comparative Law Review* 12 (1989): 193.

97. Gao and Hardin, "Export Control Risks."

98. Wrubel, "Toshiba-Kongsberg," 247.

99. "Submarined by Japan and Norway," *New York Times*, June 22, 1987, http://www .nytimes.com/1987/06/22/opinion/submarined-by-japan-and-norway.html.

100. Kelly, "Curbing Illegal Transfers," 184.

101. Department of Defense Appropriations Act, 1998, House Amendment 295. http:// thomas.loc.gov/cgi-bin/bdquery/z?d105:HZ00295.

102. Much of the work in the next section originally appeared at Robert Farley, "A Dual-Use Dilemma in US-China Defense Industrial Interaction," *The Diplomat*, November 15, 2015, http://thediplomat.com/2015/11/a-dual-use-dilemma-in-us -china-defense-industrial-interaction/.

103. Paul Mozur and Jane Perlez, "US Tech Giants May Blue National Security Boundaries in China Deals," *New York Times*, October 30, 2015. http://www.nytimes.com /2015/10/31/technology/us-tech-giants-may-blur-national-security-boundaries -in-china-deals.html.

104. Blue Heron, *Open Power, Hidden Dangers: IBM Partnerships in China* (Center for Intelligence and Research Analysis, 2015).

105. Gao and Hardin, "Export Control Risks."

106. Paul Mozur, "Taiwan Semiconductor Manufacturing to Open Own Plant in China," *New York Times*, December 7, 2015, http://www.nytimes.com/2015/12/08 /business/international/taiwan-semiconductor-china-factory-chips.html?smid= tw-share.

107. Susan K. Sell, *Private Power, Public Law: The Globalization of Intellectual Property Rights* (Cambridge University Press, 2003).

108. See, for example, Marcus Weisgerber, "The Stolen Soviet Tech in SOCOM's New Missile," *Defense One*, August 18, 2016, http://www.defenseone.com/business /2016/08/socoms-new-soviet-based-missile-deep-learning-supercomputers -defense-spending-boost-predicted-and-some-more/130873/.

109. Brooks, *Producing Security*, 104.

110. Sell, "Intellectual Property Protection and Antitrust in the Developing World," 318.

111. "Defense Patent Is an Important Part of National Patent," *Shuanglin Forging*, March 3, 2011, http://www.sxshuanglin.com/news_detail/newsId=3705481b-82b5 −4bce-a0c2–9df75f63c35e&comp_stats=comp-FrontNews_list01–1303183207302 .html.

Chapter 6

1. Von Hardesty, "Made in the USSR," *Air and Space Magazine*, March 2001, http:// www.airspacemag.com/military-aviation/made-in-the-ussr-38442437/?no-ist.

2. Hardesty, "Made in the USSR."

3. Perry McCoy Smith, *The Air Force Plans for Peace, 1943–1945* (Johns Hopkins University Press, 1970).

4. Richard Rhodes, *The Making of the Atomic Bomb* (Simon & Schuster, 1986), 605.

5. James S. Corum, "Airpower Thought in Continental Europe between the Wars," in Philip S. Meilinger, *The Paths Of Heaven: The Evolution Of Airpower Theory* (School of Advanced Airpower Studies, 2000).

6. Richard Overy, *The Bombing War: Europe, 1939–1945* (Penguin UK, 2013), 201.

7. Hardesty, "Made in the USSR."

8. Hardesty, "Made in the USSR."

9. Hardesty, "Made in the USSR."

10. Hardesty, "Made in the USSR."

11. Hardesty, "Made in the USSR."

12. Hardesty, "Made in the USSR."

13. Hardesty, "Made in the USSR."

14. Hardesty, "Made in the USSR."

15. Alexandre Carriço, *The Aviation Industry Corporation of China (AVIC) and the Research and Development Programme of the J-20* Janus.net 2, no. 2 (2011): 96–109.

16. John Harris, *Industrial Espionage and Technology Transfer: Britain and France in the Eighteenth Century* (Ashgate, 1988), 7.

17. John Harris, *Industrial Espionage*, 7.

18. Susan Sell, "Intellectual Property and Public Policy in Historical Perspective: Contestation and Settlement." *Loyola Law School Law Review* 38 (2004): 282.

19. Philip Hanson, "Soviet Industrial Espionage," *Bulletin of the Atomic Scientists* 43 no. 3, (April 1987): 25–29.

20. Larkins Dsouza, "PAKDA a Russian Stealth Bomber?" *Defense Aviation*, July 6, 2008, http://www.defenceaviation.com/2008/07/pakda-a-russian-stealth-bomber.html.

21. Planeman, "Myth Dispelled: Tu-160 Blackjack= B-1B Lancer," March 17, 2006, http://www.abovetopsecret.com/forum/thread199317/pg1.

22. One important exception to this includes the appropriation and incorporation of Soviet missile technology. See Marcus Weisberger, *The Stolen Soviet Tech in SOCOM's New Missile*, Defense One, August 18, 2016, https://www.defenseone.com/business/2016/08/socoms-new-soviet-based-missile-deep-learning-supercomputers-defense-spending-boost-predicted-and-some-more/130873/.

23. David E. Hoffman, *The Billion Dollar Spy: A True Story of Cold War Espionage and Betrayal* (New York: Doubleday, 2015).

24. David E. Hoffman, *Billion Dollar Spy*,

25. David E Sanger, "With Spy Charges, U.S. Draws a Line That Few Others Recognize," *New York Times*, May 19, 2014), http://www.nytimes.com/2014/05/20/us/us-treads-fine-line-in-fighting-chinese-espionage.html.

26. E. J. Chikofsky and J. H. Cross II" "Reverse Engineering and Design Recovery: A Taxonomy," *IEEE Software* 7 no. 1 (1990): 13–17.

27. Wendell Minnick, "Experts: China Still Lags West in Advanced Aircraft Technologies," *Defense News*, August 3, 2014, http://archive.defensenews.com/article/20140803/DEFREG03/308030011/Experts-China-Still-Lags-West-Advanced-Aircraft-Technologies.

28. Stuart Macdonald, "Nothing Either Good or Bad: Industrial Espionage and Technology Transfer," *International Journal of Technology Management* 8, no. 1–2 (1993): 95–105.

29. Andrea Gilli and Mauro Gilli, "The Diffusion of Drone Warfare? Industrial, Organizational, and Infrastructural Constraints," *Security Studies* 25 no. 1 (2016): 50–84.

30. John R. Lindsay and Tai Ming Cheung, "From Exploitation to Innovation: Acquisition, Absorption, and Application," in Jon R. Lindsay, Tai Ming Cheung, and Derek S. Reveron, eds., *China and Cybersecurity: Espionage, Strategy, and Politics in the Digital Domain* (Oxford University Press, 2015), 55.

31. See, for example,Evan S. Medeiros, Roger Cliff, Keith Crane, and James C. Mulvenon, *A New Direction for China's Defense Industry* (Rand Corporation, 2005), 68.

32. Douglas M. O'Reagan, *Taking Nazi Technology: Allied Exploitation of German Sci-*

ence after the Second World War, Kindle edition (Johns Hopkins University Press, 2019), 371.

33. Benjamin Wittes, "Other Unclassified Databases the Chinese Are Probably Stealing," Lawfare, July 27, 2015, https://www.lawfareblog.com/other-unclassified-data bases-chinese-are-probably-stealing.

34. Wittes, "Other Unclassified Databases."

35. Paul Rosenzweig and Benjamin Wittes, "Users Weigh in on What Database the PLA Should Hack Next," Lawfare, July 31, 2015, https://www.lawfareblog.com /users-weigh-what-database-pla-should-hack-next.

36. Carl Roper, *Trade Secret Theft, Industrial Espionage, and the China Threat* (CRC Press, 2014), 197.

37. Mandiant Intelligence Center, "APT1: Exposing One of China's Cyber Espionage Units," Mandiant.com (2013), 20.

38. Anonymous senior partner, interview with author.

39. David J. Betz and Tim Stevens, *Chapter One: Power and Cyberspace* (Adelphi, 2011), 35–54. See also Fritz Machlup, *The Production and Distribution of Knowledge in the United States* (Princeton University Press, 1962); Peter F. Drucker, *The Age of Discontinuity: Guidelines to Our Changing Society* (Pan Books, 1971); Daniel Bell, *The Coming of the Post-Industrial Society: A Venture in Social Forecasting* (Heinemann Educational, 1974); and Christian Fuchs, *Internet and Society: Social Theory in the Information Age* (Routledge, 2008).

40. Alex Abella, *Soldiers of Reason: The RAND Corporation and the Rise of American Empire*, Kindle edition (Harcourt, 2008), 213–15.

41. David Leigh, "How 250,000 US Embassy Cables Were Leaked," *Guardian*, November 28, 2010, http://www.theguardian.com/world/2010/nov/28/how-us-embassy -cables-leaked.

42. James Bamford, "The Most Wanted Man in the World," *Wired*, August 2014, http:// www.wired.com/2014/08/edward-snowden/.

43. Charles W. L. Hill, "Digital Piracy: Causes, Consequences, and Strategic Responses," *Asia Pacific Journal of Management* 24, no. 1 (2007): 9–25.

44. Singer, *Cybersecurity and Cyberwar*, 19–20.

45. Singer, *Cybersecurity and Cyberwar*, 19–20.

46. Brandon Valeriano and Ryan Maness, *Cyber War versus Cyber Realities: Cyber Conflict in the International System* (Oxford University Press, 2015), 22.

47. Abraham M. Denmark and Robert Kaplan, eds., *Contested Commons: The Future of American Power in a Multipolar World* (Washington: Center for New American Security, 2010); see in particular Greg Rattray, Chris Evans, and Jason Healey, "American Security in the Cyber Commons," pp. 137–76.

48. Barry R. Posen, "Command of the Commons: The Military Foundation of US Hegemony," *International Security* 28, no. 1 (Summer 2003): 8.

49. Milton Mueller, *"Ruling the Root: Internet Governance and the Taming of Cyberspace* (MIT Press, 2002). See also Mark A. Lemley, "Law and Economics of Internet Norms," *Chicago Kent Law Review* 73 (1997): 1257; Pamela Samuelson, "Intellectual Property and the Digital Economy: Why the Anti-Circumvention Regulations Need to Be Revised," *Berkeley Technology Law Journal* 14 (1999): 519.

50. Kaspersky, "What Is Spear Phishing? Definition," http://usa.kaspersky.com/inter net-security-center/definitions/spear-phishing#.VoWGRhUrLIU.

51. See, for example, Peter W. Singer and Allan Friedman, *Cybersecurity and Cyberwar: What Everyone Needs to Know* (Oxford University Press, 2014); Chris C. Demchak, *Wars of Disruption and Resilience: Cybered Conflict, Power, and National Security* (University of Georgia Press, 2011).

52. Singer and Friedman, *Cybersecurity and Cyberwar*, 17–18.

53. Singer and Friedman, *Cybersecurity and Cyberwar*, 21.

54. Singer and Friedman, *Cybersecurity and Cyberwar*, 49.

55. Singer and Friedman, *Cybersecurity and Cyberwar*, 9.

56. Singer and Friedman, *Cybersecurity and Cyberwar*, 67.

57. Singer and Friedman, *Cybersecurity and Cyberwar*, 67–69.

58. Valeriano and Maness, *Cyber War versus Cyber Realities*, 34.

59. Valeriano and Maness, *Cyber War versus Cyber Realities*," 36; Kenneth Geers, "The Challenge of Cyber Attack Deterrence." *Computer Law & Security Review* 26, no. 3 (2010): 298–303.

60. Mandiant, "APT1," 40.

61. Danny Vinik, "Survey: What Keeps America's Computer Experts Up at Night," Politico, December 9, 2015 http://www.politico.com/agenda/story/2015/12/us -cyber-attack-security-policy-survey-000338.

62. For an example of how this debate played out in the United Kingdom, see National Archives, "The Role of the Air Force in Relation to the Army: Memorandum by the Secretary of State for War," May 26, 1921. AIR 9/5 Plans Archives, vol. 47, Separate Air Force Controversy 1917–1936.

63. James Stavridis and David Weinstein, "Time for a U.S. Cyber Force," *Proceedings*, January 2014, https://www.usni.org/magazines/proceedings/2014–01/time-us-cyber-force.

64. White House, *Administration Strategy on Mitigating the Theft of U.S. Trade Secrets* (Washington: US Government Publishing Office, 2013).

65. Michael Daniel, Tony Scott, and Ed Felten, "The President's National Cybersecurity Plan: What You Need to Know," White House, February 9, 2016, https://www .whitehouse.gov/blog/2016/02/09/presidents-national-cybersecurity-plan -what-you-need-know.

66. Paul Rosenzweig, "President Obama's National Security Plan," Lawfare, February 9, 2016, https://www.lawfareblog.com/president-obamas-national-cyberse curity-plan.

67. National Cyber Strategy of the United States of America, September 2018, https:// www.whitehouse.gov/wp-content/uploads/2018/09/National-Cyber-Strategy .pdf, ii.

68. National Cyber Strategy, 1–2.

69. Ash Carter, "Rewiring the Pentagon: Charting a New Path on Innovation and Cybersecurity," Department of Defense, April 23, 2015, http://www.defense.gov /Speeches/Speech.aspx?SpeechID=1935.

70. Phillip Ewing, "Ash Carter's Appeal to Silicon Valley: We're 'Cool' Too," Politico,

April 23, 2015, http://www.politico.com/story/2015/04/ash-carter-silicon-valley-appeal-117293.html.

71. Ewing, "Ash Carter's Appeal."

72. Some of the material in this section previously appeared at Robert Farley, "A Dual-Use Dilemma in US-China Industrial Interaction," *The Diplomat*, November 2, 2015, http://thediplomat.com/2015/11/a-dual-use-dilemma-in-us-china-defense-industrial-interaction/.

73. Aimin Yan and Barbara Gray, "Bargaining Power, Management Control, and Performance in United States–China Joint Ventures: A Comparative Case Study," *Academy of Management Journal* 37, no. 6 (1994): 1478–1517.

74. "Boeing in China," Boeing, October 2015, http://www.boeing.com/resources/boeingdotcom/company/key_orgs/boeing-international/pdf/chinabackgrounder.pdf.

75. Paul Mozur and Jane Perlez, "US Tech Giants May Blue National Security Boundaries in China Deals," *New York Times*, October 30, 2015, http://www.nytimes.com/2015/10/31/technology/us-tech-giants-may-blur-national-security-boundaries-in-china-deals.html. See also Blue Heron, *Open Power, Hidden Dangers: IBM Partnerships in China* (Vienna, VA: Center for Intelligence Research and Analysis, 2015).

76. Joseph Marks, "Our Best Frenemy," Politico, December 9, 2015, http://www.politico.com/agenda/story/2015/12/china-us-cyber-attack-hacks-000332.

77. Marks, "Our Best Frenemy."

78. Lindsay and Cheung, "From Exploitation to Innovation," 66.

79. Lindsay and Cheung, "From Exploitation to Innovation," 66.

80. Xue Lin and Rocci Luppicini, "Socio-Technical Influences of Cyber Espionage: A Case Study of the GhostNet System," in Rocco Luppicini, ed., *Moral, Ethical, and Social Dilemmas in the Age of Technology: Theories and Practice* (IGI Global, 2013), 112–24.

81. Lindsay and Cheung, "From Exploitation to Innovation," 61–63.

82. Mandiant, "APT1," 2.

83. Mandiant, "APT1," 3.

84. Mandiant, "APT1," 20.

85. Joseph Marks, "Our Best Frenemy," Politico, December 9, 2015. http://www.politico.com/agenda/story/2015/12/china-us-cyber-attack-hacks-000332#ixzz3u1plBGXB.

86. Bree Feng, "Among Snowden Leaks, Details of Chinese Cyberespionage," *New York Times*, January 20, 2015, http://sinosphere.blogs.nytimes.com/2015/01/20/among-snowden-leaks-details-of-chinese-cyberespionage/.

87. Philip Dorling, "China Stole Plans for a New Fighter Plane, Spy Documents Have Revealed," *Sydney Morning Herald*, January 18, 2015, http://www.smh.com.au/national/china-stole-plans-for-a-new-fighter-plane-spy-documents-have-revealed-20150118-12sp1o.html.

88. National Security Agency, "Chinese Exfiltrate Sensitive Military Technology," *Der Spiegel*, http://www.spiegel.de/media/media-35687.pdf.

89. Steve Davies, *Red Eagles: America's Secret MiGs* (Osprey, 2008).

90. Dave Majumdar, "U.S. Pilots Say New Chinese Stealth Fighter Could Become Equal of F-22, F-35," United States Naval Institute, November 5, 2014, http://news .usni.org/2014/11/05/u-s-pilots-say-new-chinese-stealth-fighter-become-equal -f-22-f-35.

91. Michael Raska, "Strategic Contours of China's Arms Exports," RSIS Commentaries, no. 165 (2017), https://dr.ntu.edu.sg/bitstream/handle/10220/43730/CO1 7165.pdf?sequence=1&isAllowed=y/.

92. National Security Agency, "Chinese Exfiltrate Sensitive Military Technology," *Der Spiegel*, http://www.spiegel.de/media/media-35687.pdf.

93. James Vincent, "Schematics from Israel's Iron Dome Missile Shield 'Hacked' by Chinese, Says Report," *Independent*, July 29, 2014, http://www.independent.co .uk/life-style/gadgets-and-tech/israels-iron-dome-missile-shield-hacked-by -chinese-military-hackers-says-report-9635619.html.

94. Lindsay and Cheung, "From Exploitation to Innovation."

95. David Sanger and Tim Weiner. "Emerging Role for the C.I.A.: Economic Spy," *New York Times*, October 15, 1995, http://www.nytimes.com/1995/10/15/world /emerging-role-for-the-cia-economic-spy.html.

96. Sanger and Weiner, "Emerging Role."

97. Lana Lam, "Edward Snowden: US Government Has Been Hacking Hong Kong and China for Years," *South China Morning Post*, June 14, 2013, http://www.scmp.com /news/hong-kong/article/1259508/edward-snowden-us-government-has-been -hacking-hong-kong-and-china.

98. Glenn Greenwald, Ewen MacAskill, Laura Poitras, Spencer Ackerman, and Dominic Rushe, "Microsoft Handed the NSA Access to Encrypted Messages," *Guardian*, July 12, 2013, http://www.theguardian.com/world/2013/jul/11/micro soft-nsa-collaboration-user-data.

99. Shane Harris, "Exclusive: Inside the FBI's Fight against Chinese Cyber-Espionage," *Foreign Policy*, May 27, 2014, http://foreignpolicy.com/2014/05/27/exclusive-in side-the-fbis-fight-against-chinese-cyber-espionage/?wp_login_redirect=0.

100. US Department of Justice, "U.S. Charges Five Chinese Military Hackers for Cyber Espionage against U.S. Corporations and a Labor Organization for Commercial Advantage," May 19, 2014, http://www.justice.gov/opa/pr/us-charges-five-chinese -military-hackers-cyber-espionage-against-us-corporations-and-labor.

101. Paul Rosenzweig, "Executive Order on Cyber Sanctions," Lawfare, April 1, 2015, https://www.lawfareblog.com/executive-order-cyber-sanctions.

102. Shannon Tiezzi, "New US Cyber Order Could Provoke Chinese Retaliation," *The Diplomat*, April 3, 2015, http://thediplomat.com/2015/04/new-us-cyber-order -could-provoke-chinese-retaliation/.

103. Benjamin Wittes, "James Lewis on the China Cyber Deal," Lawfare, October 5, 2015, https://www.lawfareblog.com/james-lewis-china-cyber-deal.

104. Benjamin Wittes, "Maybe Those Chinese Cyber Espionage Indictments Weren't So Dumb," Lawfare, December 1, 2015, https://www.lawfareblog.com/maybe-tho se-chinese-cyber-espionage-indictments-werent-so-dumb.

105. Ellen Nakashima, "Following U.S. indictments, China Shifts Commercial Hack-

ing Away from Military to Civilian Agency," *Washington Post*, November 30, 2015, https://www.washingtonpost.com/world/national-security/following-us -indictments-chinese-military-scaled-back-hacks-on-american-industry/2015 /11/30/fcdb097a-9450–11e5-b5e4–279b4501e8a6_story.html.

106. Cody M. Poplin, "China Claims OPM Hack Was 'Criminal'; Arrests Hackers It Says Were Responsible," *Washington Post*, December 2, 2015, https://www.lawfareblog .com/china-claims-opm-hack-was-criminal-arrests-hackers-it-says-were -responsible.

107. Michael Brown and Pavneet Singh, *China's Technology Transfer Strategy* (DIUx, January 2018), 3. See also Patrick Tucker, "This Pentagon Paper Explains Why the Trump Administration Is Reining In Tech Trade with China," Defense One, April 6, 2018, https://www.defenseone.com/technology/2018/04/pentagon-pap er-explains-why-trump-administration-reining-tech-trade-china/147258/.

108. "Huawei Has Been Cut Off from American Technology," *Economist*, May 25, 2019, https://www.economist.com/business/2019/05/25/huawei-has-been-cut-off -from-american-technology.

109. Valeriano and Maness, *Cyber War versus Cyber Realities*, 22.

110. Michael N. Schmitt and Liis Vihul, "Proxy Wars in Cyberspace," *Fletcher Security Review* 1, no. 2 (Spring 2014), 55–56.

111. Sanger, "With Spy Chargese."

112. Kevin McCauley, "PLA Transformation: Difficult Military Reforms Begin," *China Brief* 15, no. 18 (September 9, 2015). http://www.jamestown.org/single/?tx_ttnews %5Btt_news%5D=44349&tx_ttnews%5BbackPid%5D=7&cHash=36fc86d567f57 eef0ebfe8ead43e6ea9#.VoUa3hUrLIU.

113. Matthew Dahl, "What Effect Could Chinese Military Reorganization Have on the Recent US-China Cyber Agreement?" *Lawfare*, November 3, 2015, https:// www.lawfareblog.com/what-effect-could-chinese-military-reorganization-have -recent-us-china-cyber-agreement.

114. Jim Garamone, "U.S. Cyber Command Chief Details Plans to Meet Cyber-space Threats," *DoD News*, September 8, 2015, http://www.defense.gov/News -Article-View/Article/616512/us-cyber-command-chief-details-plans-to-meet -cyberspace-threats.

115. Andrew Tilghman, "Does Cyber Corps Merit Its Own Service Branch?" *Military Times*, August 21, 2015, http://www.militarytimes.com/story/military/pentagon /2015/04/09/cyber-corps-merit-own-service-branch/25530133/.

116. Marks, "Our Best Frenemy."

117. Marks, "Our Best Frenemy."

118. Fergus Hanson, "Norms of Cyber War in Peacetime," Lawfare, November 15, 2015, https://www.lawfareblog.com/norms-cyber-war-peacetime.

119. Graham Webster, "Has U.S. Cyber Pressure Worked on China?," *The Diplomat*, December 10, 2015. http://thediplomat.com/2015/12/has-u-s-cyber-pressure-wo rked-on-china/.

120. Marks, "Our Best Frenemy."

121. Lucas Kello, "The Meaning of the Cyber Revolution: Perils to Theory and State-craft." *International Security* 38, no. 2 (2013): 7–40.

Chapter 7

1. Mark Mazzeti, *The Way of the Knife: The CIA, A Secret Army, and a War at the Ends of the Earth* (Penguin, 2013), 66.

2. Some of this material previously appeared in a similar form in Robert Farley, *Grounded: The Case for Abolishing the United States Air Force*, Kindle edition (University Press of Kentucky, 2014) 2616–52.

3. Thomas P. Ehrhard, *Air Force UAVs: The Secret History* (Defense Technical Information Center, 2010), 4.

4. Ehrhard, *Air Force UAVs*, 4.

5. Ehrhard, *Air Force UAVs*, 5.

6. Ehrhard, *Air Force UAVs*, 13.

7. Ehrhard, *Air Force UAVs*, 35.

8. Ehrhard, *Air Force UAVs*, 40.

9. Ehrhard, *Air Force UAVs*, 49.

10. Sharon Weinberger, *The Imagineers of War: The Untold Story of DARPA, the Pentagon Agency That Changed the World* (Knopf, 2017), 264.

11. Lee H. Kean and Thomas H. Hamilton, *The 9/11 Report* (St. Martin's, 2004), 16, 213–14.

12. Federation of American Scientists, "Tactical Common Datalink," June 21, 1997, http://fas.org/irp/program/disseminate/tcdl.htm.

13. Hasik, *Arms and Innovation*, 576.

14. Hasik, *Arms and Innovation*, 488.

15. Parts of the following appeared in similar form in Robert Farley, "Did China's Military Drone Technology Espionage Pay Off in the End?" *The Diplomat*, February 19, 2016, http://thediplomat.com/2016/02/did-chinas-military-drone-technology-espionage-pay-off-in-the-end/.

16. Robert Johnson, "China's Mysterious Predator Clone Is Finally Out in the Open," *Business Insider*, November 8, 2012, http://www.businessinsider.com/chinas-mysterious-predator-clone-is-finally-out-in-the-open-2012–11.

17. Bill Gertz, "China's Armed Drones Appear Built from Stolen Data from US Cyber Intrusions," *Asia Times*, December 29, 2015, http://atimes.com/2015/12/chinas-armed-drones-appear-built-from-stolen-data-from-us-cyber-intrusions/.

18. Gertz, "China's Armed Drones."

19. Kyle Mizokami, "Turns Out Buying a Chinese Knock-Off Predator Drone Is a Bad Idea," *Popular Mechanics*, June 12, 2019 https://www.popularmechanics.com/military/aviation/a27926078/predator-drone-chinese/.

20. Greg Waldron, "China Finds Its UAV Export Sweet Spot," *FlightGlobal*, June 14, 2019. https://www.flightglobal.com/military-uavs/china-finds-its-uav-export-sweet-spot/132557.article.

21. Susan K. Sell, *Private Power, Public Law: The Globalization of Intellectual Property Rights* (Cambridge University Press, 2003), 80.

22. Ministry of Commerce, "Intellectual Property Protection in China," http://www.chinaipr.gov.cn/.

23. David Shambaugh, *China Goes Global: The Partial Power* (Oxford University Press, 2013), 131.

24. Henry Farrell and Abraham Newman, "Weaponized Interdependence," *International Security* 44, no. 1 (2019).

25. Sell, *Private Power, Public Law.*

26. Greg Austin, "The Problem with China's Patents," *The Diplomat*, March 3, 2015, http://thediplomat.com/2015/03/the-problem-with-chinas-patents/.

27. Ministry of Commerce of the People's Republic of China, "Intellectual Property Protection in China," http://www.chinaipr.gov.cn/.

28. Wang Fan, "PLA Opens Door to Invite Civil Firms in Defense-Tech R&D," ECNS .cn, November 26, 2014, http://www.ecns.cn/military/2014/11–26/144270.shtml.

29. Wayne M. Morrison, "Enforcing U.S. Trade Laws: Section 301 and China," Congressional Research Service, June 11, 2019, https://fas.org/sgp/crs/row/IF10708 .pdf.

30. Edward Hallett Carr, *The Twenty Years' Crisis, 1919–1939: An Introduction to the Study of International Relations* (Harper & Row, 1964).

SELECTED BIBLIOGRAPHY

Abella, Alex. *Soldiers of Reason: The RAND Corporation and the Rise of American Empire*. Harcourt, 2008.

Adamsky, Dima. *The Culture of National Innovation: The Impact of Cultural Factors on the Revolution in Military Affairs in Russia, the US, and Israel*. Stanford University Press, 2010.

Aizenman, Joshua, and Reuven Glick (2006). "Military Expenditure, Threats, and Growth." *Journal of International Trade & Economic Development* 15, no. 2 (2006): 129–55.

Alderman, Daniel Lisa Crawford, Brian Lafferty, and Aaron Shraberg. "The Rise of Chinese Civil-Military Integration." In Tai Ming Cheung, ed., *Forging China's Military Might: A New Framework for Assessing Innovation*: 109–35.

Alford, William P. *To Steal a Book Is an Elegant Offense: Intellectual Property Law in Chinese Civilization*. Stanford University Press, 1995.

Alic, John A. *Beyond Spinoff: Military and Commercial Technologies in a Changing World*. Harvard Business Press, 1992.

Allison, John R., and Lianlian Lin. "Evolution of Chinese Attitudes toward Property Rights in Invention and Discovery." *University of Pennsylvania Journal of International Economic Law* 20 (1999): 735–92.

Anderson, Jennifer. "The Limits of Sino-Russian Partnership." Adelphi Paper 315 (1997).

Anthony, Ian. "Economic Dimensions of Soviet and Russian Arms Exports." In Ian Anthony, ed., *Russia and the Arms Trade*, 71–92. Oxford University Press, 1998.

Arms Control Association. "International Code of Conduct against Ballistic Missile Proliferation." https://www.armscontrol.org/documents/icoc.

Avant, Deborah D. *Political Institutions and Military Change: Lessons from Peripheral Wars*. Cornell University Press, 1994.

Basberg, Bjørn L. "Patents and the Measurement of Technological Change: S Survey of the Literature." *Research Policy* 16, no. 2 (1987): 131–41.

Bell, Daniel. *The Coming of the Post-Industrial Society: A Venture in Social Forecasting*. Heinemann Educational, 1974.

Bellany, Ian. "The Offensive-Defensive Distinction, the International Arms Trade, and Richardson and Dewey." *Peace and Conflict: Journal of Peace Psychology* 1, no. 1 (1995): 37–48.

Berman, Harold J., and John R. Garson. "United States Export Controls: Past, Present, and Future." *Columbia Law Review* 67, no. 5 (1967): 791–890.

Biddle, Stephen. *Military Power: Explaining Victory and Defeat in Modern Battle*. Princeton University Press, 2004.

Black, Jeremy. *Naval Warfare: A Global History since 1860*. Rowman & Littlefield, 2017.

Blue Heron. *Open Power, Hidden Dangers: IBM Partnerships in China*. Center for Intelligence and Research Analysis, 2015.

Bracken, Paul, Linda Brandt, and Stuart E. Johnson, "The Changing Landscape of Defense Innovation," *Defense Horizons* 47 (July 2005): 1–8.

Brooks, Stephen G. *Producing Security: Multinational Corporations, Globalization, and the Changing Calculus of Conflict*. Princeton University Press, 2007.

Brown, Michael, and Pavneet Singh. *China's Technology Transfer Strategy*. Defense Innovation Unit, January 2018.

Budylin, Sergey, and Yulia Osipova. "Total Upgrade: Intellectual Property Law Reform in Russia," *Columbia Journal of East European Law* 1, no. 1 (2007): 1–39.

Burk, Dan L., and Mark A. Lemley. "Is Patent Law Technology-Specific?" *Berkeley Technology Law Journal* 17 (2002): 1155–1206.

———. "Policy Levers in Patent Law." *Virginia Law Review* (2003): 1575–1696.

Carr, Edward Hallett. *The Twenty Years' Crisis, 1919–1939: An Introduction to the Study of International Relations*. Harper & Row, 1964.

Carriço, Alexandre. "The Aviation Industry Corporation of China (AVIC) and the Research and Development Programme of the J-20." Janus.net 2, no. 2, (2011): 96–109.

Carter, Ash. "Rewiring the Pentagon: Charting a New Path on Innovation and Cybersecurity." Department of Defense, April 23, 2015. http://www.defense.gov/Speeches/Speech.aspx?SpeechID=1935.

Cassman, Daniel R. "Keep It Secret, Keep It Safe: An Empirical Analysis of the State Secrets Doctrine." *Stanford Law Review* 67 (2015): 1173–1216.

Cervano, Francis. "Export Controls and Their Relationship to National Defense Industries." In Richard A. Bitzinger, ed., *The Modern Defense Industry: Political, Economic, and Technological Issues*, 243–56. ABC-CLIO, 2009.

Chandler, Scott E. *Rethinking Competition in Defense Acquisition*. Lexington Institute, 2014.

Chari, P. R. "Indo-Soviet Military Cooperation: A Review." *Asian Survey* 19, no. 3 (1979): 230–44.

Chayes, Abram, and Antonia Handler Chayes. "On Compliance." *International Organization* 47, no. 2 (1993): 175–205.

Chesney, Robert M. "State Secrets and the Limits of National Security Litigation," *George Washington Law Review* 75 (2006): 1249–1332.

Cheung, Tai Ming. "The Chinese Defense Economy's Long March from Imitation to Innovation." In Tai Ming Cheung, ed., *China's Emergence as a Defense Technological Power*, 31–60. Routledge, 2013.

———. *Fortifying China: The Struggle to Build a Modern Defense Economy*. Cornell University Press, 2013.

————. "An Uncertain Tradition." In Tai Ming Cheung, ed., *Forging China's Military Might: A New Framework for Assessing Innovation*, 47–65. John Hopkins University Press, 2014.

Cheung, Tai Ming, Thomas G. Mahnken, and Andrew L. Ross. "Frameworks for Analyzing Chinese Defense and Military Innovation." In Tai Ming Cheung, ed., *Forging China's Military Might: A New Framework for Assessing Innovation*, 666–73. Johns Hopkins University Press, 2014.

Chikofsky, E. J., and J. H. Cross II. "Reverse Engineering and Design Recovery: A Taxonomy." *IEEE Software* 7, no. 1 (1990): 13–17.

Chivers, C. J. *The Gun*. Simon & Schuster, 2010.

Clapham, Andrew. *Human Rights Obligations of Non-State actors*. Oxford University Press, 2006.

Coleman, Katharina P., and Michael W. Doyle. "Introduction: Expanding Norms, Lagging Compliance." In Edward C. Luck and Michael W. Doyle, eds., *International Law and Organization: Closing the Compliance Gap*. Rowman and Littlefield, 2004.

Collins, Gabe, and Andrew Erickson. "Jet Engine Development in China: Indigenous High-Performance Turbofans are a Final Step toward Fully Independent Fighter Production." *China SignPost*, June 26, 2011, http://www.chinasignpost.com /2011/06/26/jet-engine-development-in-china-indigenous-high-performance -turbofans-are-a-final-step-toward-fully-independent-fighter-production/.

Cook, Lisa D. "A Green Light for Red Patents? Evidence from Soviet Domestic and Foreign Inventive Activity, 1962 to 1991." Working paper, 2010.

Crane, Keith, Roger Cliff, Evan Medeiros, James Mulvenon, and William Overholt. *Modernizing China's Military: Opportunities and Constraints*. RAND, 2005.

Cutler, Robert M., Laure Després, and Aaron Karp. "The Political Economy of East-South Military Transfers." *International Studies Quarterly* 31, no. 3 (September 1987): 273–99.

Daniels, Mario. "Restricting the Transnational Movement of 'Knowledgeable Bodies': The Interplay of US Visa Restrictions and Export Controls in the Cold War." In John Krige, ed., *How Knowledge Moves: Writing the Transnational History of Science and Technology*, 35–64. University of Chicago Press, 2019.

Davies, Robert William, Mark Harrison, and Stephen G. Wheatcroft. *The Economic Transformation of the Soviet Union, 1913–1945*. Cambridge University Press, 1994.

Davies, Steve. *Red Eagles: America's Secret MiGs*. Osprey, 2008.

Demchak, Chris C. *Wars of Disruption and Resilience: Cybered Conflict, Power, and National Security*. University of Georgia Press, 2011.

Desch, Michael C. "Don't Worship at the Altar of Andrew Marshall." *National Interest*, December 17, 2014. http://nationalinterest.org/feature/the-church-st-andy-11867.

Deutsch, John. "Consolidation of the US defense industrial base." *Acquisition Review Quarterly* 8, no. 3 (2001): 138–50.

DiMaggio, Paul J., and Walter W. Powell. "The Iron Cage Revisited: Institutional Isomorphism and Collective Rationality in Organizational Fields." *American Sociological Review* (1983): 147–60.

Dombrowski, Peter J., and Eugene Gholz. *Buying Military Transformation: Technological Innovation and the Defense Industry.* Columbia University Press, 2006.

Dombrowski, Peter, and Andrew Ross. "The Revolution in Military Affairs, Transformation, and the U.S. Defense Industry." In Richard A. Bitzinger, ed., *The Modern Defense Industry: Political, Economic, and Technological Issues: Political, Economic, and Technological Issues*, 153–74. ABC-CLIO, 2009.

Douglas, Susan J. *Inventing American broadcasting 1899–1922.* Johns Hopkins Univ. Pr., 1987.

Drahos, Peter and John Braithwaite. "Who Owns the Knowledge Economy?: Political Organising Behind TRIPS." Corner House Briefing No. 32, http://www .thecornerhouse.org.uk/item.shtml?x=85821 (Sept. 2004).

———. *Information Feudalism: Who Owns the Knowledge economy?* New Press, 2007).

Drexl, Josef. "Intellectual Property and Implementation of Recent Bilateral Trade Agreements in the EU." In Josef Drexl, Henning Grosse Ruse-Khan, and Souheir Nadde-Phlix, eds,. *EU Bilateral Trade Agreements and Intellectual Property: For Better Or Worse?* Springer, 2014: 265–291.

Drezner, Daniel W. "Globalization and Policy Convergence." *International Studies Review* 3, no. 1 (2001): 53–78.

Drucker, Peter F. *The Age of Discontinuity: Guidelines to Our Changing Society.* Pan Books, 1971.

Dunne, Paul J. "Developments in the Global Arms Industry from the End of the Cold War to the mid-2000s." In Richard A. Bitzinger, ed. *The Modern Defense Industry: Political, Economic, and Technological Issues: Political, Economic, and Technological Issues*, 13–37. ABC-CLIO, 2009.

Edquist, Charles. *Systems of Innovation: Technologies, Institutions, and Organizations.* Pinter, 1997.

Ehrhard, Thomas P. *Air Force UAVs: The Secret History.* Defense Technical Information Center, 2010.

Eikenberry, Karl W. *Explaining and Influencing Chinese Arms Transfers.* No. 36. DIANE Publishing, 1995.

Epstein, Katherine C. "Intellectual property and national security: the case of the hardcastle superheater, 1905–1927." *History and Technology* 34, no. 2 (2018): 126–156.

———. *Torpedo.* Harvard University Press, 2014).

Etzion, Dror, and Gerald F. Davis. "Revolving Doors? A Network Analysis of Corporate Officers and US Government Officials." *Journal of Management Inquiry* 17, no. 3 (2008): 157–61.

Etzioni, Amitai. "International Prestige, Competition and Peaceful Coexistence." *European Journal of Sociology* 3, no. 1 (1962): 21–41.

Evans, David C., and Mark R. Peattie. *Kaigun: Strategy, Tactics, and Technology in the Imperial Japanese Navy 1887–1941.* Seaforth, 2012.

Eyer, Dana P., and Mark C. Suchman. "Status, Norms, and the Proliferation of Conventional Weapons: An Institutional Theory Approach." In Peter J. Katzenstein, ed., *The Culture of National Security*, 79–113. Columbia University Press, 1996.

Farley, Robert. "Making the PLAN: A Concise History of the People's Liberation Army Navy." In Donovan C. Chau and Thomas Kane, eds., *China and International Security*, 99–118. ABC Clio, 2014.

———. *Grounded: The Case for Abolishing the United States Air Force*. University Press of Kentucky, 2014.

Farrell, Henry, and Abraham Newman. "Weaponized Interdependence." *International Security* 44, no. 1 (2019).

Farrell, Theo. "World Culture and the Irish Army, 1922–1942." In Theo Farrell and Terry Terriff, eds., *The Sources of Military Change: Culture, Politics, Technology*, 69–90. Lynne Rienner Publishers, 2002.

Federico, Pasquale Joseph. "Origin and Early History of Patents." *Journal of the Patent and Trademark Office Society* 11 (1929): 292–305.

Feiveson, Harold A., and Jacqueline W. Shire. "Dilemmas of Compliance with Arms Control and Disarmament Agreements." In Edward C. Luck and Michael W. Doyle, eds., *International Law and Organization: Closing the Compliance Gap*. Rowman and Littlefield, 2004.

Finnegan, Phillip. "The Evolution of Defense Hierarchies." In Richard A. Bitzinger, ed., *The Modern Defense Industry: Political, Economic, and Technological Issues: Political, Economic, and Technological Issues*, 95–106. ABC-CLIO, 2009.

Finnemore, Martha. *The Purpose of Intervention: Changing Beliefs about the Use of Force*. Cornell University Press, 2003.

Finnemore, Martha, and Kathryn Sikkink. "International Norm Dynamics and Political Change." *International Organization* 52, no. 0 (1998): 887–917.

Fiott, Daniel. "Europe and the Pentagon's Third Offset Strategy." *RUSI Journal* 161, no. 1 (2016): 26–31.

Fisher, Louis. "The State Secrets Privilege: Relying on 'Reynolds.'" *Political Science Quarterly* 122, no. 3 (2007): 385–408.

Fosfuri, Andrea, and Thomas Rønde. "High-Tech Clusters, Technology Spillovers, and Trade Secret Laws." *International Journal of Industrial Organization* 22, no. 1 (2004): 47–48.

Fuchs, Christian. *Internet and Society: Social Theory in the Information Age*. Routledge, 2008.

Galtsova, Paulina. "Intellectual Property Reform in Russia: Analysis of Part Four of the Russian Civil Code." MA thesis, Lund University Faculty of Law, 2008.

Gansler, Jacques S., William S. Greenwalt, and William Lucyshin. *Non-Traditional Commercial Defense Contractors*. Center for Public Policy and Private Enterprise, 2013.

Garrett, Geoffrey. "Global Markets and National Politics: Collision Course or Virtuous Circle?" *International Organization* 52, no. 4 (1998): 787–824.

Geers, Kenneth. "The Challenge of Cyber Attack Deterrence." *Computer Law & Security Review* 26, no. 3 (2010): 298–303.

Geyer, Michael. "German Strategy in the Age of Machine Warfare, 1914–1945." In Peter Paret and Gordon Craig, eds., *Makers of Modern Strategy: From Machiavelli to the Nuclear Age*, 527–97. Princeton University Press, 1986.

Gilli, Andrea, and Mauro Gilli. "The diffusion of drone warfare? Industrial, organizational, and infrastructural constraints." *Security Studies* 25, no. 1 (2016): 50–84.

———. "Why China Has Not Caught Up Yet: Military-Technological Superiority and the Limits of Imitation, Reverse Engineering, and Cyber Espionage." *International Security* 43, no. 3 (2019): 141–89.

Goldman, Emily. "The Spread of Western Military Models to Ottoman Turkey and Meiji Japan." In Theo Farrell and Terry Terriff, eds., *The Sources of Military Change: Culture, Politics, Technology*, 41–68. Lynne Riener, 2002.

Goldman, Emily O., and Leslie C. Eliason, eds. *The Diffusion of Military Technology and Ideas*. Stanford University Press, 2003.

Goldstein, Judith. "International Law and Domestic Institutions: Reconciling North American "Unfair" Trade Laws." *International Organization* 50, no. 4 (1996): 541–64.

Gorman, G. Scott. "The TU-4: The Travails of Technology Transfer by Imitation." *Air Power History* 45, no. 1 (1998): 16–27.

Gormley, Dennis M. *Missile Contagion: Cruise Missile Proliferation and the Threat to International Security*. Naval Institute Press, 2008.

Goure, Daniel. *Incentivizing a New Defense Industrial Base*. Lexington Institute, 2015.

Gourevitch, Peter. "The Second Image Reversed: The International Sources of Domestic Politics." *International Organization* 32, no. 4 (1978): 881–912.

Gray, Colin. *Weapons Don't Make War: Policy, Strategy, and Military Technology*. University Press of Kansas, 1993.

Gray, Peter. *Leadership, Direction and Legitimacy of the RAF Bomber Offensive from Inception to 1945*. Continuum, 2012.

Haas, Ernst B. *Beyond the Nation-State: Functionalism and International Organization*. Stanford University Press, 1964.

Habeck, Mary. *Storm of Steel: The Development of Armor Doctrine in Germany and the Soviet Union, 1919–1939*. Cornell University Press, 2003.

Hannas, William C., James Mulvenon, and Anna B. Puglisi. *Chinese Industrial Espionage: Technology Acquisition and Military Modernization*. Routledge, 2013.

Hanson, Philip. "Soviet Industrial Espionage." *Bulletin of the Atomic Scientists* 43, no. 3 (1987): 25–29.

Hardesty, Von "Made in the USSR." *Air and Space Magazine*, March 2001. http://www.airspacemag.com/military-aviation/made-in-the-ussr-38442437/?no-ist.

Harris, John. *Industrial Espionage and Technology Transfer: Britain and France in the Eighteenth Century*. Ashgate, 1988.

Harrison, Mark. "The Soviet Union: The Defeated Victor." In Mark Harrison, ed., *The Economics of World War II: Six Great Powers in International Comparison*, 268–301. Cambridge University Press, 2000.

Hasik, James M. *Arms and Innovation: Entrepreneurship and Alliances in the Twenty-First-Century Defense Industry*. University of Chicago Press, 2008.

———. "Better Buying Power or Better Off Not? Purchasing Technical Data for Weapon Systems." *Defense Acquisition Research Journal* 21, no. 3 (July 2014): 694–714.

Hayward, Keith. "The Globalization of Defense Industries." In Richard A. Bitzinger, ed., *The Modern Defense Industry: Political, Economic, and Technological Issues: Political, Economic, and Technological Issues,* 107–22. ABC-CLIO, 2009.

Henk, Daniel W., and Marin Revavi Rupiya. *Funding Defense: Challenges of Buying Military Capability in Sub-Saharan Africa.* Strategic Studies Institute, 2001.

Hicks, Katherine, et al. "Assessing the Third Offset Strategy." Center for Strategic and International Studies, March 2017. https://csis-prod.s3.amazonaws.com/s3fs -public/publication/170302_Ellman_ThirdOffsetStrategySummary_Web.pdf.

Hill, Alexander. *The Red Army and the Second World War.* Cambridge University Press, 2016.

Hill, Charles W. L. "Digital Piracy: Causes, Consequences, and Strategic Responses." *Asia Pacific Journal of Management* 24, no. 1 (2007): 9–25.

Hoffman, David E. *The Billion Dollar Spy: A True Story of Cold War Espionage and Betrayal.* Doubleday, 2015.

Holbrook, Timothy R. "Extraterritoriality in US Patent Law." *William & Mary Law Review* 49 (2007): 2119–98.

Horn, Henrik, Petros C Mavroidis, and Andre Sapir. "Beyond the WTO? An Anatomy of EU and US Preferential Trade Agreements." CEPR Discussion Paper No. DP7317 (June 2009). https://ssrn.com/abstract=1433913.

Horowitz, Michael. *The Diffusion of Military Power: Causes and Consequences for International Politics.* Princeton University Press, 2010.

Horowitz, Richard S. "Beyond the Marble Boat: The Transformation of the Chinese Military, 1850–1911," in David A. Graff and Robin Higham, eds., *A Military History of China,* 153–74. Westview Press, 2002.

Howard, Michael, and Peter Paret, eds. *Carl von Clausewitz on War.* Princeton University Press, 1984.

Howells, Jeremy. "Tacit Knowledge, Innovation and Technology Transfer." *Technology Analysis & Strategic Management* 8, no. 2 (1996): 91–106.

Hu, Albert Guangzhou, and Gary H. Jefferson. "Returns to Research and Development in Chinese Industry: Evidence from State-Owned enterprises in Beijing." *China Economic Review* 15, no. 1 (2004): 86–107.

Hu, Albert Guangzhou and Ivan P. Png. "Patent Rights and Economic Growth: Evidence from Cross-Country Panels of Manufacturing Industries." CELS 2009 4th Annual Conference on Empirical Legal Studies paper, August 10, 2012. https:// www.wipo.int/edocs/mdocs/mdocs/en/wipo_ip_econ_ge_5_10/wipo_ip_econ _ge_5_10_ref_huandpng.pdf.

Huber, George P. "Organizational Learning: The Contributing Processes and the Literatures." *Organization Science* 2, no. 1 (1991): 88–115.

Isaacs, Davida H. "Not All Property Is Created Equal: Why Modern Courts Resist Applying the Takings Clause to Patents, and Why They Are Right to Do So." *George Mason Law Review* 15 (2007): 1–44.

Isaacs, Davida H., and Robert M. Farley. "Privilege-Wise and Patent (and Trade Secret) Foolish? How the Courts' Misapplication of the Military and State Secrets Privilege Violates the Constitution and Endangers National Security." *Berkeley Technology Law Journal* 24, no. 2 (2009): 785–818.

Jakobsen, Linda, Paul Holtom, Dean Knox, and Jingchao Peng. *China's Energy and Security Relations with Russia: Hopes, Frustrations, Uncertainties.* SIPRI, October 2011.

Jervis, Robert. "Cooperation under the Security Dilemma." *World Politics* 30, no. 2 (January 1978): 167–214.

Johns, Adrian. Piracy: *The Intellectual Property Wars from Gutenberg to Gates.* University of Chicago Press, 2010.

Karp, Aaron. "The Global Small Arms Industry: Transformed by War and Society." In Richard A. Bitzinger, ed., *The Modern Defense Industry: Political, Economic, and Technological Issues: Political, Economic, and Technological Issues,* 272–92. ABC-CLIO, 2009.

Kean, Lee H., and Thomas H. Hamilton. *The 9/11 Report.* St. Martin's, 2004.

Kello, Lucas. "The Meaning of the Cyber Revolution: Perils to Theory and Statecraft." *International Security* 38, no. 2 (2013): 7–40.

Kelly, Bryan, Dimitris Papanikolaou, Amit Seru, and Matt Taddy, "Measuring Technological Innovation over the Long Run." National Bureau of Economic Research working paper no. 25266 (2018). https://www.nber.org/papers/w25266.

Kelly, Stephen D. "Curbing Illegal Transfers of Foreign-Developed Critical High Technology from CoCom Nations to the Soviet Union: An Analysis of the Toshiba-Kongsberg Incident." *Boston College International and Comparative Law Review* 12 (1989): 181–224.

Kier, Elizabeth. "Culture and Military Doctrine: France Between the Wars." *International Security* 19, no. 4 (Spring 1995): 65–93.

———. *Imagining War: French and British Military Doctrine between the Wars.* Princeton University Press, 1997.

Kirshin, Yuriy. "Conventional Arms Transfers during the Soviet Period." In Ian Anthony, ed., *Russia and the Arms Trade,* 38–70. Oxford University Press, 1998.

Koh, Harold, Hongju Abram Chayes, Antonia Handler Chayes, and Thomas M. Franck. "Why Do Nations Obey International Law?" *Yale Law Journal* 106, no. 8 (1997): 2599–2659.

Kortunov, Sergey. "The Influence of External Factors on Russia's Arms Export Policy." In Ian Anthony, ed., *Russia and the Arms Trade,* 93–106. Oxford University Press, 1998.

Krepinevich, Andrew F. "Cavalry to Computer: The Pattern of Military Revolutions." *The National Interest* 37 (Fall 1994): 30–42.

Krepinevich, Andrew, and Barry Watts. *The Last Warrior: Andrew Marshall and the Shaping of Modern American Defense Strategy.* Basic Books, 2015.

Krige, John. "Export Controls as Instruments to Regulate Knowledge Acquisition in a Globalizing Economy." In John Krige, ed., *How Knowledge Moves: Writing the Transnational History of Science and Technology,* 62–92. University of Chicago Press, 2019.

Lai, Edwin L.-C., "International Intellectual Property Rights Protection and the Rate of Product Innovation." *Journal of Development Economics* 55, no. 1 (1998): 133–53.

Larson, Daniel. "Yesterday's Technology, Tomorrow: How the Government's Treatment of Intellectual Property Prevents Soldiers from Receiving the Best Tools to Complete Their Mission." *John Marshall Review of Intellectual Property Law* 7, no. 171 (2007): 171–204.

Lee, Jaewon. "South Korea's Export Control System." Stockholm International Peace Research Institute, November 2013. http://books.sipri.org/files/misc/SIPRIBP1311 .pdf.

Lee, Sabing H. "Protecting the Private Inventor under the Peacetime Provisions of the Invention Secrecy Act." *Berkeley Technology Law Journal* 12 (1997): 345–412.

Leiber, Keir. *War and the Engineers: The Primacy of Politics over Technology.* Cornell University Press, 2005.

Leigh, David. "How 250,000 US Embassy Cables Were Leaked." *Guardian,* November 28, 2010. http://www.theguardian.com/world/2010/nov/28/how-us-embassy -cables-leaked.

Lieber, Keir A. "Grasping the Technological Peace: The Offense-Defense Balance and International Security." *International Security* 25, no. 1 (Summer 2000): 71–104.

Lin, Xue, and Rocci Luppicini. "Socio-Technical Influences of Cyber Espionage: A Case Study of the GhostNet System." In Rocci Luppicini, ed., *Moral, Ethical, and Social Dilemmas in the Age of Technology: Theories and Practice: Theories and Practice.* IGI Global, 2013.

Lindsay, John R., and Tai Ming Cheung. "From Exploitation to Innovation: Acquisition, Absorption, and Application." In Jon R. Lindsay, Tai Ming Cheung, and Derek S. Reveron, eds., *China and Cybersecurity: Espionage, Strategy, and Politics in the Digital Domain,* 51–86. Oxford University Press, 2015.

Lobell, Steven E. "Second Image Reversed Politics: Britain's Choice of Freer Trade or Imperial Preferences, 1903–1906, 1917–1923, 1930–1932." *International Studies Quarterly* 43, no. 4 (1999): 671–93.

Lofgren, Charles A. "United States v. Curtiss-Wright Export Corporation: An Historical Reassessment." *Yale Law Journal* 83, no. 1 (1973): 1–32.

Luck, Edward C. "Gaps, Commitments, and the Compliance Challenge." In Edward C. Luck and Michael W. Doyle, eds., *International Law and Organization: Closing the Compliance Gap.* Rowman and Littlefield, 2004.

Lundvall, Bengt-Ake, ed. *National Systems of Innovation: Towards a Theory of Innovation and Interactive Learning.* Pinter, 1992.

Lynn-Jones, Sean M. "Offense-Defense Theory and Its Critics." *Security Studies* 4, no. 4 (Summer 1995): 660–91.

Macdonald, Stuart. "Nothing Either Good or Bad: Industrial Espionage and Technology Transfer." *International Journal of Technology Management* 8, no. 1–2 (1993): 95–105.

Machlup, Fritz. *The Production and Distribution of Knowledge in the United States.* Princeton University Press, 1962.

Mandiant Intelligence Center. "APT1: Exposing One of China's Cyber Espionage Units." Mandiant.com, 2013.

Martens, John A. *Secret Patenting in the USSR and Russia.* Deep North Press, 2010.

Mathieu, Charlotte. "Assessing Russia's Space Cooperation with China and India:

Opportunities and Challenges for Europe." *Acta Astronautica* 66, no. 3–4 (2010): 355–61.

Matthews, Ron, and Curie Maharani. "The Defense Iron Triangle Revisited." In Richard A. Bitzinger, ed., *The Modern Defense Industry: Political, Economic, and Technological Issues: Political, Economic, and Technological Issues*, 38–60. ABC-CLIO, 2009.

Mazzeti, Mark. *The Way of the Knife: The CIA, A Secret Army, and a War at the Ends of the Earth*. Penguin, 2013.

McAllen, Dorothy K. "National Security Policy Constraints on Technological Innovation: A Case Study of the Invention Secrecy Act of 1951." Eastern Michigan University master's theses and doctoral dissertations, paper 580, 2012.

McCauley, Kevin. "PLA Transformation: Difficult Military Reforms Begin," Jamestown Foundation, China brief volume 15, no. 18, September 9, 2015. http://www.jamestown.org/single/?tx_ttnews%5Btt_news%5D=44349&tx_ttnews%5BbackPid%5D=7&cHash=36fc86d567f57eef0ebfe8ead43e6ea9#.VoUa3hUrLIU.

Medeiros, Evan S., Roger Cliff, Keith Crane, and James C. Mulvenon. *A New Direction for China's Defense Industry*. RAND, 2005.

Mehrotra, Santosh. *India and the Soviet Union: Trade and Technology Transfer*. Cambridge University Press, 1990.

Merges, Robert. "Battle of the Lateralisms: Intellectual Property and Trade." *Boston University International Law Journal* 8, no. 2 (Fall 1990): 245.

Meyer, John W., John Boli, George M. Thomas, and Francisco O. Ramirez. "World Society and the Nation-State." *American Journal of Sociology* 103, no. 1 (July 1997): 144–81.

Ministry of Commerce of the People's Republic of China. "Intellectual Property Protection in China." http://www.chinaipr.gov.cn/.

Moon, Chung-in. "South Korea and International Compliance Behavior: The WTO and IMF in Comparative Perspective." In Edward C. Luck and Michael W. Doyle, eds., *International Law and Organization: Closing the Compliance Gap*. Rowman and Littlefield, 2004.

Moran, Theodore H. "The Globalization of America's Defense Industries: Managing the Threat of Foreign Ddependence." *International Security* 15, no. 1 (1990): 57–99.

Morrison, Wayne M. "Enforcing U.S. Trade Laws: Section 301 and China." Congressional Research Service, June 11, 2019. https://fas.org/sgp/crs/row/IF10708.pdf.

Mueller, Milton. *Ruling the root: Internet governance and the taming of cyberspace*. MIT press, 2002.

Murray, Williamson and Allan R. Millett eds. *Military Innovation in the Interwar Period*. Cambridge University Press, 1996.

Neild, Robert. "Defining" Offensive:" A Failure and a Success." *Bulletin of the Atomic Scientists* 44, no. 7 (1988): 18.

Nelson, Richard R., ed. *National Innovation Systems: A Comparative Analysis* (Oxford University Press, 1993).

Neuman, Stephanie G. "Power, Influence, and Hierarchy: Defense Industries in a

Unipolar World." In Richard A. Bitzinger, ed., *The Modern Defense Industry: Political, Economic, and Technological Issues: Political, Economic, and Technological Issues,* 60–94. ABC-CLIO, 2009.

Newman, Lily Hay. "Who Owns the Software in the Car You Bought?" *Slate Magazine,* May 22, 2015. http://www.slate.com/blogs/future_tense/2015/05/22/gm_and_john_deere_say_they_still_own_the_software_in_cars_customers_buy.html.

Nordhaus, William D. "An Economic Theory of Technological Change." *American Economic Review* 59, no. 2 (1969): 18–28.

———. "The Optimum Life of a Patent: Reply." *American Economic Review* 62, no. 3 (1972): 428–31.

Office of the US Trade Representative. "Chile Free Trade Agreement." https://ustr.gov/trade-agreements/free-trade-agreements/chile-fta.

———. "Intellectual Property Rights in U.S.-South Korea Trade Agreement." https://ustr.gov/uskoreaFTA/IPR.

———. "United States–Korea Free Trade Agreement: Final Text." https://ustr.gov/trade-agreements/free-trade-agreements/korus-fta/final-text.

O'Reagan, Douglas M. *Taking Nazi Technology: Allied Exploitation of German Science after the Second World War.* Johns Hopkins University Press, 2019.

Overy, Richard. *The Bombing War: Europe, 1939–1945.* Penguin UK, 2013).

Paarlberg, Robert L. "Knowledge as Power: Science, Military Dominance, and US Security." *International Security* 29, no. 1 (2004): 122–51.

Pagano, Ugo. "The Crisis of Intellectual Monopoly Capitalism." *Cambridge Journal of Economics* 38, no. 6 (2014): 1409–29.

Paul, T. V. "Influence through Arms Transfers: Lessons from the US-Pakistani Relationship." *Asian Survey* 32, no. 12 (1992): 1078–92.

Pooley, James. "Trade Secrets: The Other Intellectual Property Right." *WIPO Magazine,* June 2013. https://www.wipo.int/wipo_magazine/en/2013/03/article_0001.html.

Posen, Barry R. "Command of the Commons: The Military Foundation of US Hegemony." *International Security* 28, no. 1 (2003): 5–46.

———. *The Sources of Military Doctrine: France, Britain, and Germany between the World Wars.* Cornell University Press, 1984.

Pridham, David. "Steal a Trade Secret, Go to Jail?" *Forbes,* June 1, 2017. https://www.forbes.com/sites/davidpridham/2017/06/01/steal-a-trade-secret-go-to-jail/#683fcbaa6a47.

Purcell, Edward A. "Understanding Curtiss-Wright." *Law and History Review* 31, no. 4 (2013): 653–715.

Quester, George H. *Offense and Defense in the International System.* Wiley, 1977.

Raska, Michael. "Strategic Contours of China's Arms Exports." RSIS Commentaries, no. 165 (2017). https://dr.ntu.edu.sg/bitstream/handle/10220/43730/CO17165.pdf?sequence=1&isAllowed=y/.

Resende-Santos, João. *Neorealism, States, and the Modern Mass Army.* Cambridge University Press, 2007.

Rhodes, Richard. *The Making of the Atomic Bomb.* Simon & Schuster, 1986.

Rogowski, Ronald. *Commerce and Coalitions: How Trade Affects Domestic Political Alignments*. Princeton University Press, 1989.

Rohwer, Jürgen, and Mikhail S. Monakov. *Stalin's Ocean-Going Fleet: Soviet Naval Strategy and Shipbuilding Programmes, 1935–1953*. Psychology Press, 2001.

Roper, Carl. *Trade Secret Theft, Industrial Espionage, and the China Threat*. CRC Press, 2014.

Rosen, Stephen Peter. *Societies and Military Power: India and Its Armies*. Cornell University Press, 1996.

———. *Winning the Next War: Innovation and the Modern Military*. Cornell University Press, 1991.

Rousseau, Richard. "The Torturous Sino-Russian Arms Trade-Analysis." *Eurasia Review*, June 9, 2012, http://www.eurasiareview.com/09062012-the-tortuous-sino-russian-arms-trade-analysis/.

Ryan, Michael Patrick. *Knowledge Diplomacy: Global Competition and the Politics of Intellectual Property*. Brookings Institution Press, 1998.

Sagan, Scott. *The Limits of Safety: Organizations, Accidents, and Nuclear Weapons*. Princeton University Press, 1993.

Sakakibara, Mariko, and Lee Branstetter. Do Stronger Patents Induce More Innovation? Evidence from the 1988 Japanese Patent Law Reforms. Working paper no. w7066, National Bureau of Economic Research, 1999.

Sanger, David E. "With Spy Charges, U.S. Draws a Line That Few Others Recognize." *New York Times*, May 19, 2014. http://www.nytimes.com/2014/05/20/us/us-treads-fine-line-in-fighting-chinese-espionage.html.

Sanger, David, and Tim Weiner. "Emerging Role for the C.I.A.: Economic Spy." *New York Times*, October 15, 1995. http://www.nytimes.com/1995/10/15/world/emerging-role-for-the-cia-economic-spy.html.

Schmitt, Michael N., and Liis Vihul. "Proxy Wars in Cyberspace." *Fletcher Security Review* 1, no. 2 (Spring 2014): 53–72.

Schmooker, Jacob. "Economic Sources of Inventive Activity." *Journal of Economic History* 22, no. 1 (1962): 1–20.

Schweller, Randall L. "Domestic Structure and Preventive War: Are Democracies More Pacific?" *World Politics* 44, no. 2 (1992): 235–69.

Sebenius, James K., and Rebecca Hulse. "Charlene Barshefsky: HBS 9–801–421." *Harvard Business Review* (2013).

Sell, Susan. "Intellectual Property and Public Policy in Historical Perspective: Contestation and Settlement." *Loyola of Los Angeles Law Review* 38 (2004): 267–322.

———. *Private Power, Public Law: The Globalization of Intellectual Property Rights*. Cambridge University Press, 2003.

Shambaugh, David. *China Goes Global: The Partial Power*. Oxford University Press, 2013.

Sibley, Katherine, and Amelia Siobhan. *Red Spies in America: Stolen Secrets and the Dawn of the Cold War*. University Press of Kansas, 2004.

Singer, Peter W. and Allan Friedman. *Cybersecurity and Cyberwar: What Everyone Needs to Know*. Oxford University Press, 2014.

SIPRI. "Arms Transfers Database." https://www.sipri.org/databases/armstransfers.

Smith, Perry McCoy. *The Air Force Plans for Peace, 1943–1945*. Johns Hopkins University Press, 1970.

Solis, Gary D. *The Law of Armed Conflict: International Humanitarian Law in War*. Cambridge University Press, 2010.

Sontag, Sherry, Christopher Drew, and Annette Lawrence Drew. *Blind Man's Bluff: The Untold Story of Cold War Submarine Espionage*, 158. Random House, 2000.

Starr, Paul. *The Creation of the Media: Political Origins of Modern Communications*. Basic Books, 2004.

Stavridis, James, and David Weinstein. "Time for a U.S. Cyber Force." *Proceedings*, January 2014. https://www.usni.org/magazines/proceedings/2014–01/time-us -cyber-force.

Stohl, Rachel, and Suzette Grillot. *International Arms Trade*. Wiley, 2013.

Sweet, Alec S., Wayne Sandholtz, and Neil Fligstein. *The Institutionalization of Europe*. Oxford University Press, 2001.

Tallberg, Jonas. "Paths to Compliance: Enforcement, Management, and the European Union." *International Organization* 56, no. 3 (2002): 609–43.

Teece, David J. "Technology Transfer by Multinational Firms: The Resource Cost of Transferring Technological Know-How." *Economic Journal* 87, no. 346 (1977): 242–61.

Tomes, Robert R. "Relearning Counterinsurgency Warfare." *Parameters* 34, no. 1 (spring 2004): 16–28.

Tyroler-Cooper, Samm, and Alison Peet. "The Chinese Aviation Industry: Techno-Hybrid Patterns of Development in the C919 Program." In Tai Ming Cheung, ed., *China's Emergence as a Defense Technological Power*, 89–110. Routledge, 2013.

Valeriano, Brandon and Ryan Maness. *Cyber War versus Cyber Realities: Cyber Conflict in the International System*. Oxford University Press, 2015.

Varun, Sahni. "Not Quite British: A Study of External Influences on the Argentine Navy." *Journal of Latin American Studies* 25, no. 3 (1993): 489–513.

Waltz, Kenneth. *Man, the State, and War: A Theoretical Analysis*. Columbia University Press, 1959.

———. *Theory of International Politics*. McGraw-Hill, 1979.

Wang, Jian-Ye, and Magnus Blomström. "Foreign Investment and Technology Transfer: A Simple Model." *European Economic Review* 36, no. 1 (1992): 137–55.

Watts, Barry D. *The Evolution of Precision Strike*. Center for Strategic and Budgetary Assessments, 2013.

Weinberger, Sharon. *The Imagineers of War: The Untold History of DARPA, the Pentagon Agency that Changed the World*. Knopf, 2017.

White House. "National Cyber Strategy of the United States of America." September 2018. https://www.whitehouse.gov/wp-content/uploads/2018/09/National-Cyber -Strategy.pdf.

———. *Administration Strategy on Mitigating the Theft of U.S. Trade Secrets*. US Government Publishing Office, 2013.

Wrubel, Wende A. "Toshiba-Kongsberg Incident: Shortcomings of Cocom, and Recommendations for Increased Effectiveness of Export Controls to the East Bloc." *American University Journal of International Law and Policy* 4 (1989): 221–73.

Yan, Aimin, and Barbara Gray. "Bargaining Power, Management Control, and Performance in United States–China Joint Ventures: A Comparative Case Study." *Academy of Management Journal* 37, no. 6 (1994): 1478–1517.

Yu, Peter K. "From Pirates to Partners: Protecting Intellectual Property in China in the Twenty-First Century." *American University Law Review* 50 (2001): 131–244.

Zisk, Kimberley Martin. *Engaging the Enemy: Organization Theory and Soviet Military Innovation, 1955–1991.* Princeton University Press, 1993.

Zweig, David, Chen Changgui, and Stanley Rosen. "Globalization and Transnational Human Capital: Overseas and Returnee Scholars to China." *China Quarterly* 179 (2004): 735–57.

INDEX